AF323960

Steve Huyton

Timeless Classics

Modern Dress Wristwatches

SCHIFFER PUBLISHING

4880 Lower Valley Road • Atglen, PA 19310

Library of Congress Control Number: 2024931519

Edited by Ian Robertson
Cover design by Danielle Farmer
Type set in Scotch/New Frank

ISBN: 978-0-7643-6845-5
Printed in India

Published by Schiffer Publishing, Ltd.
4880 Lower Valley Road
Atglen, PA 19310
Phone: (610) 593-1777; Fax: (610) 593-2002
Email: Info@schifferbooks.com
Web: www.schifferbooks.com

For our complete selection of fine books on this and related subjects, please visit our website at www.schifferbooks.com. You may also write for a free catalog.

Schiffer Publishing's titles are available at special discounts for bulk purchases for sales promotions or premiums. Special editions, including personalized covers, corporate imprints, and excerpts, can be created in large quantities for special needs. For more information, contact the publisher.

We are always looking for people to write books on new and related subjects. If you have an idea for a book, please contact us at proposals@schifferbooks.com.

Timeless Classics

Modern Dress Wristwatches

VACHERON CONSTANTIN
GENEVE

Contents

TRILOBE
SWISS MADE

Foreword

by Bernhard Lederer

A wristwatch fulfills many tasks and functions. It is primarily an instrument for displaying the current time. It can be a useful tool in dealing with small and large periods of time, or an impressive demonstration of what is technically feasible. It is actually a miniaturized image of the cosmos.

The interrelationships between space and time have always fascinated watchmakers and inspired them to achieve top technical achievements. Measuring the course of the sun and dividing the days into equally long portions was soon no longer enough for the versatile Zeit mechanics. They began to decipher the constellations that determine the phases of the moon and the length of the year. The accuracy of their models always depended on the precise rate of the clockwork, so even the most accurate calendars were only approximate attempts to depict reality. The rule over precision has always been incumbent on the divine celestial machine, the mode of operation of which we barely understand.

Modern digital technology has broken down the gigantic world mechanics into the smallest bits and bytes and linked them with mathematical models to calculate the passage of time in advance and compare it with the measurement of the decay of atomic nuclei. No matter how precise it is and wherever it is and always available, it does not give rise to a feeling for the passage of time, for the preciousness of the moment or the value of a heartbeat. In mechanical timekeeping, on the other hand, one addresses these questions with constructive understanding, technical skill, and aesthetic sense. And watchmakers—probably in contrast to mathematicians—humbly realize that it is not they who set the pace of the time.

In recent years, technically complicated or at least complicated-looking timepieces have dominated watch fashion, but a countercurrent has now been established that wants to return to the essence of the time display. The watches should be simple and functional, elegant and refined mainly in the lines, the choice of materials, and the surface treatment. The technical effort to provide the exact time is not displayed here. Nowadays this is most visible through the glazed case back and can thus also be admired in its aesthetics.

The modern dress watch thus reflects the natural quality hierarchy of the watch. In the heyday of modern mechanics, the value of a wristwatch was defined by two factors: accuracy and movement height. The more precise it was or the flatter it was, the more expensive it was. This resulted in an ideal of beauty that has basically endured to this day, characterized by flat housings, flowing shapes, and stylistic minimalism.

The clockworks do their job hidden under the dial. The fact that these are downright miracles of micromechanics of the very highest processing quality can be seen at any time from the precision of the time display. And judge.

Viewed in this way, the crowning glory of horological creation is in reality not the tourbillon, which ostensibly looks great, and not the perpetual calendar, but the dress watch, the epitome of the elegant wristwatch, a discreet calling card of personal taste and an expression of individual style.

Introduction

Over the last decade the world of horology has changed dramatically. Brands like DeBethune MB & F and Urwerk achieved almost rock star status for their bold, innovative designs. At that time, oversized asymmetrical cases were in vogue and more contemporary materials were filtering into the industry. Nowadays numerous avant-garde watchmakers are experimenting with different time displays that break away from the conventional three-hand indication.

On a personal level, I absolutely love massively oversized modernistic watches made from lightweight industrial materials. Even though these timepieces have a bold aesthetic, sometimes telling the time is a challenge. Another factor, if you have smaller wrists, is that massive watches can look too overbearing. They also are unsuitable for formal events because they protrude over the sleeve of a dinner jacket. This is a reason why dress watches have such a mass appeal, because, basically, they are practical.

For my university dissertation I wrote about Cary Grant and his contribution to light comedy. In my opinion he was the epitome of sophistication and style. In these classical movies, Cary Grant would wear bespoke designer suits accompanied by an exquisite dress watch. At this time there was a limited selection to choose from. However, today there is a huge variety of different designs available on the market. This book explores a diverse list of exciting options from a broad spectrum of brands.

In the world of horology, many trends are transient, but dress-style watches have always appealed to a broad demographic. This book features numerous modern interpretations from this genre. As well as selections from the large groups (LVMH, Richemont, Swatch) there are examples from many small microbrands. In the spirit of diversity I wanted to include quite a few affordable options. However, every timepiece featured in the book is based entirely on merit and aesthetic value.

Modern Dress Watches
A. Lange & Söhne

A. Lange & Söhne is a prestigious watch brand that was originally founded by
Ferdinand Adolph Lange in 1845. The company's headquarters are in Glashütte,
Germany, and since 2000 have been a subsidiary of the Richemont Group.

Little Lang 1 Moon Phase

As beautiful as a starry night. Thousands of tiny stars seem to sparkle on the dark-blue dial. Essentially this is how A. Lange & Söhne describes the Little Lang 1 Moon Phase. This watch is ideal for formal occasions and has an extravagant 18-karat 36.8mm × 10mm white-gold case. Impressive features include a white-gold moon indication, iconic Lange oversized date, and raised Roman numerals. At the heart of the watch is a refined movement that showcases nine different artisan techniques, including artistic decorations, polishing, and engraving. This 44-jewel mechanism contains 411 individual components and oscillates at a frequency of 21,600 vibrations per hour.

Triple Split

The Triple Split is the world's only mechanical chronograph that allows multihour comparative measurements in scores for applications like automobile races or equestrian sports. This beautiful dress watch is presented in an exquisite 43.2mm × 15.6mm 18-karat pink-gold case and is limited to only 100 pieces. What makes this timepiece so desirable is the sublime blue dial with applied gold baton indexes. Functionally the watch features hours, minutes, seconds, chronograph with 12-hour split seconds, and power reserve indicator. Beneath the luxurious façade is a sophisticated 46-jewel mechanism that contains 567 components and oscillates at a frequency of 21,600 vibrations per hour.

Aerowatch

Aerowatch prides itself on being a reputable family-owned watch brand. The company has been operational since 1910 and is in the picturesque Jura mountain region of Saignelégier, Switzerland.

7 TIME ZONES

7 Time Zones is one of Aerowatch's flagship models from their stylish renaissance collection. What makes this watch so special is the highly distinctive dial featuring seven different time zones. The timepiece showcases local time in London, Moscow, Dubai, Tokyo, Shanghai, and New York. These are laid out on six individual black circular satin-finished counters. Other refined characteristics include central hour/minute hands, a domed sapphire crystal, and a pusher at 10 o'clock. Beneath the 44mm steel case lies a modified Unitas 6497 movement comprising 17 jewels and oscillating at a frequency of 18,000 vibrations per hour.

AEROWATCH
SWISS   MADE

Skeleton Spider

The Skeleton Spider is a new addition to the highly popular renaissance collection from Aerowatch. This model features a spider skeletonized matte black dial that has a refined contemporary aesthetic. Key details include a second hand graced with a spider, satin red chapter ring, a domed sapphire crystal, and luminous indexes. The timepiece also has a high-quality, 43mm black PVD-coated steel case. Powering the watch is a manual winding customized and skeletonized Unitas movement that comprises 17 jewels and oscillates at a frequency of 18,000 vibrations per hour.

Andersen Genève

Genève is the brand name of industry veteran Sven Andersen, who founded the AHCI (Académie Horlogère des Créateurs Indépendants) in 1985. The company is in the heart of Geneva, Switzerland.

Jumping Hour 40th Anniversary

The conception of the Jumping Hour dates to 1995, when a distinguished Spanish collector asked Svend Andersen to create a special unique 1/1 piece. Subsequently other editions like the "Grand Jour & Nuit" Edition were launched. In 2020 ANDERSEN Genève created a special version to commemorate his 40th anniversary. This timepiece is presented in a luxurious 38mm 5N red-gold case. What makes this watch so amazing is the hand guilloché 21-karat BlueGold "Magical Losanges" dial, which was exceptionally challenging to manufacture. Other refined features include a jumping hours indication displayed via a window (12 o'clock) and small minute counter (6 o'clock). Powering the watch is a bespoke Frédéric Piguet double-barrel automatic movement with "Jumping Hours" mechanism developed and manufactured by ANDERSEN Genève.

Worldtimer

In 2021, Andersen Genève collaborated with watch collector Benjamin Chee on a special Worldtimer project called the "Celestial Voyager." This exciting preseries version is a limited edition (seven pieces) and is encapsulated in an extravagant platinum 950 37.8mm case. For this model the brand has created sculptural polished "eagle wings" lugs. What makes this timepiece so enticing is the multipart dial with rotating ring displaying 24-hour time zones and a central disc finished in cloisonné enamel. Other features include a world time zone scale (on an Aventurine stone disc) and white-gold city names that can be personalized to the client's specifications. Beneath the sophisticated façade lies a 17-jewel self-winding movement with hand guilloché 21-karat BlueGold rotor.

Angelus Watch

Angelus was originally founded in 1891 by brothers Albert and Gustav Stoltz. After a 30-year hiatus the brand was revived in 2015 by Manufacture La Joux-Perret (Citizen Holdings) and is now associated with fine quality timepieces.

U20 Ultra-Skeleton Tourbillon

The U20 Ultra-Skeleton Tourbillon watch has a "Haute Horlogerie" style sculptural appearance. Angelus say the timepiece is "built like a supercar; its design is streamlined and based on structural optimization." Interestingly, this model is fabricated from cutting edge materials like sapphire, carbon, and titanium. However, the watch would still be suitable to wear for formal occasions. Beneath the refined façade is a highly complicated manual winding movement that features a flying tourbillon, sapphire main plate, and blue titanium bridges. The Calibre A-250 comprises 18 jewels and oscillates at a frequency of 21,600 vibrations per hour.

U30 Tourbillon Rattrapante

The U30 Tourbillon Rattrapante combines three coveted "haute horlogerie" complications in one contemporary watch: a tourbillon, fly-back double column-wheel chronograph, and split seconds (rattrapante). What makes this timepiece so desirable is the exquisite architectural skeletonized dial. This feature is encapsulated in a 47mm × 15mm grade 5 titanium case. Other refined details include a black chapter ring and 30-minute counter laser engraved onto the upper chronograph bridge. Powering the watch is a mechanical self-winding movement. The Calibre A-150 comprises 38 jewels and oscillates at a frequency of 28,800 vibrations per hour.

Angular Momentum

Angular Momentum is a luxury goods company that specializes in creating bespoke fountain pens, pocket watches, and wristwatches. The company is the brainchild of master craftsman Martin Pauli and is in Bern, Switzerland.

Tamamushi Timepiece

The Tamamushi Timepiece is a prestigious model from the Metiers d'Art Urushi collection. Interestingly, Urushi is the sap of the urushi, or lacquer tree, which is native to Japan, China, and Korea. The sap of this tree contains a resin that polymerizes and becomes a very hard, durable, plastic-like substance when it is exposed to moisture and air. Ultimately this makes it an ideal choice for the dial of a modern dress watch. For this model, Martin Pauli has chosen to decorate the dial with an ornament of small squares cut from "Tamamushi" jewel beetle wings. He has also opted to present the timepiece in a classically proportioned 42mm stainless steel case. At the heart of the watch is a Swiss-made self-winding mechanism that has been finished and assembled by hand.

Bengal Tiger

The Bengal Tiger is an exquisite dress watch that delicately combines traditional artisan techniques with traditional styling. This watch is part of the highly successful Metiers d'Art Verre Églomisé collection. Essentially, Verre Eglomisé is a term used to describe the art of cold painting and gilding on the reverse of glass. For the dial of this particular model, Martin Pauli has depicted an image of a reclining Bengal tiger. Other sophisticated highlights include a rose gold aperture at 6 o'clock displaying hours on a revolving mother-of-pearl disc. This entire splendor is encapsulated in an extravagant 41 mm rose gold case. At the heart of the watch is a historical self-winding Fontainmelon FHF 96 movement manufactured between 1950 and 1960.

Armin Strom

Armin Strom is a high-end Swiss brand that was originally established by its namesake Armin Strom in 1967. Since 2006 the company has been owned by businessman Serge Michel and master watchmaker Claude Greisler.

Gravity Equal Force

The Gravity Equal Force is the world's first automatic watch with constant force transmission. Essentially this means consistent power is delivered to the balance, ensuring the watch has flawlessly constant precision. There are several models in the range, including the extravagant Manufacture Edition Black Gold. This timepiece has a luxurious 41mm × 12.65mm 18-karat rose gold case that is ideal to wear with a tailored suit. The multilayered dial is highly effective and gives the recipient a glimpse of the mechanism. Other details include an offset subdial and rose gold hour/minute hands. Powering the watch is the caliber ASB19, which comprises 28 jewels and oscillates at a frequency of 25,200 vibrations per hour.

Tribute 1

As the name might suggest, the "Tribute 1" honors the lifetime achievement of company founder Armin Strom. The timepiece displays classical proportions in stainless steel measuring 38mm. In fact, the brand states they have redefined the classic dress watch. Features like an offset blue dial featuring hours, minutes, and seconds give the timepiece a modernistic appearance. Other details include an exposed motor barrel, an antireflective sapphire crystal, and a black alligator leather strap. Beneath the façade lies a 21-jewel mechanical hand winding movement with a 100-hour power reserve. The Calibre AMW21 is composed of 135 individual components and oscillates at a frequency of 25,200 vibrations per hour.

Arnold & Son

Luxury brand Arnold & Son is named after 18th-century English watchmaker John Arnold, who was renowned for his ingenuity and work on marine chronometers. The company is owned by and acquires all its movements from Manufacture La Joux-Perret (Citizen Holdings).

Luna Magna

The Luna Magna features Arnold & Son's first three-dimensional moon. This magnificent three-dimensional detail is fabricated from marble and Aventurine. Other highlights include an exquisite Aventurine glass façade with an off-center white lacquered hours subdial. Functionally the timepiece features hours, minutes, and astronomical three-dimensional moon phases (one-day correction required every 122 years). At the heart of the watch is a sophisticated manually winding mechanical movement. The Calibre A&S1021 comprises 35 jewels and oscillates at a frequency of 21,600 vibrations per hour.

Globetrotter Gold

Arnold & Son describes the Globetrotter Gold as a finely crafted atlas. What makes this timepiece so alluring is the central dome, which depicts the Northern Hemisphere. Every detail has been meticulously crafted, including mountain ranges and oceans hand-painted with several coats of blue-pigmented lacquer. Other refined features include a solid gold central bridge and domed sapphire crystal. Beneath the 45mm 18-karat gold case lies a highly complicated movement. The Calibre A&S6022 comprises 29 jewels and oscillates at a frequency of 28,800 vibrations per hour.

Azimuth

Neuchâtel-based watchmaker Azimuth has delighted buyers with their exceptionally well-designed timepieces since 2003. The team comprises chief product visionary Christopher Long and technical director Giuseppe Picchi.

56 BIT Singapore Commemorative Garfield Edition

At Baselworld 2009, Azimuth impressed the world with a complicated timepiece called "Back-In-Time (BIT)." Fast forward to 2021, and the company has created a special version to commemorate Singapore's 56th birthday. Interestingly, chief product visionary Christopher Long has partnered with cartoon character Garfield, adding a very playful touch. Within the range there's a steel version (56 pieces) and a very exclusive collectors' model (15 pieces). Both are equally impressive and are designed to make a bold statement. Key features include a dial depicting Garfield in a fencing outfit. Powering the watch is a Swiss mechanical automatic movement sourced from Selita. The Calibre SW220 comprises 26 jewels and oscillates at a frequency of 28,800 vibrations per hour.

Round-1 Dragon RRM

The Round-1 Dragon RRM was originally released in 2008 and was instantly popular with collectors worldwide. Recently the brand resurrected the series with a special limited edition (25 pieces). Unlike many of Azimuth's avant-garde models, this piece is ideally suited to more formal occasions. With a stainless steel case measuring 42mm × 14.8mm, it will also appeal to the unisex market. What makes this timepiece so special is the exquisite graded etched dial with bright yellow embossed dragon motif and mother-of-pearl subdial. Functionally the watch features regulator hours and retrograde minutes and seconds. Beneath the stylish façade lies an in-house-modified 23-jewel Swiss mechanical movement that oscillates at a frequency of 21,600 vibrations per hour.

Backes & Strauss

Backes & Strauss is an English brand that creates high-end timepieces and jewelry. The business dates to 1789 in Hanau, Germany, and prides itself on being the oldest diamond company in the world.

Berkeley King Tourbillon

The Berkeley King Tourbillon is a highly extravagant offering from Backes & Strauss. This watch would be an absolutely ideal choice for a glitzy function. What makes this timepiece so luxurious is the 18 karat white gold 43mm square case inlaid with 136 baguette-cut diamonds. This opulent theme continues with a white mother-of-pearl dial containing seven blue baguette-cut sapphire indexes and 116 white diamonds. In my opinion the pièce de résistance is the exposed tourbillon carriage. Powering the watch is a Swiss-made mechanical self-winding movement. This mechanism has a customized rotor engraved with an illustration of the London skyline.

Piccadilly Earl of Strauss 45 Brilliant

The Piccadilly Earl of Strauss 45 Brilliant is a modern interpretation of the classical dress watch. Dimensionally the timepiece exudes generous proportions and has a lightweight titanium case measuring 45mm. What makes this unique 1/1 piece so appealing is the highly distinctive, multi-layered, partly skeletonized dial containing 367 white diamonds. Other features include a gem-set crown and a high-quality alligator leather strap. At the heart of the watch is a mechanical movement with date and a 120-hour power reserve. As a perfect finishing touch, Backes & Strauss have meticulously set 167 white diamonds (0.75 carats) onto the bridge. This is showcased via the sapphire crystal exhibition case back.

Baume & Mercier

Baume & Mercier has a rich history dating to 1830. Over the years they have established a solid reputation for creating high-quality affordable watches. In 1988 the brand was acquired by the Richemont Group.

Baume Ocean

The Baume Ocean is a stylish dress watch with a contemporary appearance. This timepiece is a departure for the brand because the 42mm gray/blue case is fabricated in recycled plastic. The company states the watch is a collaboration with the NGO WFO and the SEAQUAL initiative. Essentially it reflects a directive in eco-responsibility and innovation. These are two key progressions in the brand's development. Other features include an interchangeable strap made of woven RPET lined with natural black cork. At the heart of the watch is a Swiss-made automatic mechanical movement made by Selita. The Calibre SW200 comprises 26 jewels and oscillates at a frequency of 28,800 vibrations per hour.

Riviera 10616

The Riviera 10616 revives a Baume & Mercier classic that was originally released in 1973. This 2021 edition has a more contemporary aesthetic consistent with modern dress watches. What makes this timepiece so special is the 42mm stainless steel case with a distinctive dodecagonal bezel (equipped with four screws). This is complemented by a matching interchangeable steel bracelet. Other refined details include a smoky-blue decorated sapphire dial, Roman numeral indexes, and open-worked hour/minute hands. Powering the watch is a Swiss-made mechanical self-winding "Baumatic" movement. The Calibre BM13-1975A comprises 21 jewels and oscillates at a frequency of 28,800 vibrations per hour.

Behrens Original

Behrens Original is a relatively new watch label based in the heart of Hong Kong. The name of the brand is a tribute to German architect and designer Peter Behrens, one of the important figures to the modernist movement.

B022 –The Rotary

B022 –The Rotary is one of Behrens' Originals flagship models. Impressively, the watch was recently nominated for the Grand Prix d'Horlogerie de Genève award. This highly prestigious organization highlights and yearly rewards the most remarkable contemporary creations and promotes the art of watchmaking worldwide. Aesthetically the timepiece has a refined avant-garde appearance reminiscent of creations by Urwerk. However, the timepiece has a smaller 42mm case that is ideal to wear with formal attire. Key details include a gray multilayered dial and dome sapphire crystal. Powering the watch is a modified Swiss automatic mechanical movement from ETA. The Calibre 2824 comprises 25 jewels and oscillates at a frequency of 28,800 vibrations per hour.

B023 – The Da Vinci Code

The B023 – The Da Vinci Code is inspired by a fictional novel of the same name. Essentially the concept focuses on a secret code hidden in a painting for more than 2,000 years. This watch has a very multilayered architectural appearance but is still suitable to wear for formal occasions. Time is displayed by an innovative satellite system composed of three revolving discs. There is also a date indication positioned at 2 o'clock. Other features include a 42mm 316L stainless steel case and domed sapphire crystal. Beneath the modern façade lies a modified Swiss automatic mechanical movement. The Calibre SW200 comprises 26 jewels and oscillates at a frequency of 28,800 vibrations per hour.

Bovet Fleurier

Bovet Fleurier SA was originally founded in London in 1822 by Swiss watchmaker Édouard Bovet. The current company is now privately owned by Pascal Raffy and specializes in creating highly exclusive timepieces.

19Thirty Red Guilloche

19Thirty Red Guilloche is a beautiful dress watch inspired by 19th-century pocket watches. According to the brand, the red of the dial symbolizes life, health, love, and passion. This timepiece is available in a choice of 42mm 18 karat red gold or stainless steel cases. What makes the watch so appealing is the meticulous hand-decorated red guilloché dial. This is an ornamental technique that has been in use since the Middle Ages. Other features include raised Roman numerals, power reserve indication, a small second subdial, and distinctive crown system. Beneath the luxurious façade lies a sophisticated mechanical hand-winding movement. The Calibre 15BM04 oscillates at 21,600 vibrations per hour and has a power reserve of seven days.

BOVET
1822
Brainstorm
Chapter Two
Phase de lune
SUD
NORD
Swiss Handcrafted
TOKYO
SYDNE
NOUMEA
FIJI
TONGA
HAWAII
ALASKA
L.ANGELE
DENVER
MEXICO
BEIJIN
BANGKok
DHAKA
KARACHI
DUBAI
MOSCOW
KIEV
LONDON
AZORES
S.GEORGIA
S.PAULO
HALIFAX
12
9
3
6

Monsieur BOVET

The Amadéo Fleurier 43 Monsieur BOVET is a delightful dress watch with a refined appearance. Interestingly, the timepiece displays the hours, minutes, and seconds on both sides. This is enabled by a patented double coaxial display. Both dials feature the ornamental guilloché technique and also exposed elements of the mechanism. This entire splendor is encapsulated in an extravagant 43mm 18 karat white or red gold case. As an additional feature the timepiece is available with a full-skin alligator strap or gold-plated silver necklace (to use as a pocket watch). Powering the watch is an in-house Swiss mechanical hand-winding movement. The Calibre 13BM09AI oscillates at 21,600 vibrations per hour and has a power reserve of seven days.

Brequet

Breguet is a luxury watch and clock manufacturer originally founded by legendary horologist Abraham-Louis Breguet in Paris in 1775. Since 1999 the brand has been a subsidiary of the Swiss Swatch Group.

Classique Double Tourbillon

The Classique Double Tourbillon is one of Brequet's most extravagant watches and part of the "Grande Complication" collection. This timepiece has a luxurious 950 platinum case measuring 46mm × 16.80mm. What makes this watch so desirable is the sensational multilayered skeletonized dial. Other distinctive features include a sapphire disc chapter ring with Roman numerals and a hand-engraved center plate. There are also two independent exposed tourbillons affixed by a bridge. Overall, the composition is very sophisticated and the finishing is superlative. At the heart of the watch is an in-house manufacture manual-winding movement. The Calibre 588 N comprises 81 jewels and contains 738 individual components.

Tradition 7097

The Tradition 7097 is an exquisite dress watch with a sophisticated contemporary appearance. Within the range there are two 40mm case variants that are available in 18-karat rose or white gold. What makes this watch so remarkable is the off-centered dial in silvered gold, which is entirely engine-turned by hand. This characteristic gives the recipient a glimpse of the anthracite mechanism, which also features retrograde seconds. Beneath the luxurious façade is a self-winding movement with silicon balance spring and a sapphire crystal case back (30 mm). The Calibre 505 SR1 comprises 38 jewels and contains 239 individual components. This timepiece also has a power reserve of 50 hours and is water-resistant to a depth of 30 meters.

Bulgari

Bulgari is a high-end luxury Italian brand that specializes in the production of fine fragrances, jewelry, leather goods, and timepieces. Currently the company is part of the LVMH (Moët Hennessy Louis Vuitton) group.

Octo Finissimo REF 102714

The Octo Finissimo REF 102714 is a striking dress watch from Bulgari. This timepiece is inspired by 16th-century Italian artists like Leonardo Da Vinci, who excelled in creating both innovative and timelessly beautiful masterpieces. What makes this watch so enticing is the 40mm octagonal case, which is finely crafted in titanium. Features like the skeletonized dial are very effective and give the watch an architectural appearance. Other fine details include a slimline chapter ring, open-worked hour/minute hands, and a petite second counter at 7 o'clock. Beneath the stylish façade is an exceptionally slimline mechanical hand-winding movement. The Calibre BVL 128SK is only 2.35mm thick and oscillates at a frequency of 28,800 vibrations per hour.

Gérald Genta Arena Bi-Retro Sport

The Gérald Genta Arena Bi-Retro Sport is a fantastic dress watch with a refined contemporary twist. This 43mm timepiece pays homage to the legendary Swiss watch designer who is famous for designing the Royal Oak (Audemars Piguet) and the Nautilus (Patek Philippe). In 2000 they bought the rights to use the Gérald Genta name. What makes this timepiece so special is the modernistic textured black dial with complementary bold yellow numerals. Other features include oversized skeletonized yellow hands that display the minute and date functions in a retrograde format. There is also a large hour window positioned at 12 o'clock. Powering the watch is a Swiss mechanical automatic movement. The Calibre BVL BVL 300 oscillates at a frequency of 28,800 vibrations per hour.

Cédric Johner

Cédric Johner is a specialist independent horologist that creates bespoke watches for discerning customers worldwide. The brand was originally established in 1997 and is in Carouge, Geneva, Switzerland.

Abyss Chronograph

The Abyss Chronograph is an exceptionally high-quality dress style watch that has a refined modernistic appearance. Dimensionally the timepiece conforms to classical proportions and has a stainless steel case measuring 41.20mm × 39mm × 11.80mm. What makes the watch stand out from the crowd is the vibrant blue grade 5 titanium dial, which achieves its color and texture through anodization. Other fine details include raised silver counters and Roman numerals. Due to the handcrafted artisan techniques used in production, each of the 30 limited pieces has a different aesthetic. Beneath the sculptural façade is a custom made mechanical movement. This caliber is fastidiously polished by hand and oscillates at a frequency of 18,000 vibrations per hour.

Maestria

The Maestria is a highly exclusive dress watch entirely handcrafted in Cédric Johner's workshop, based in Carouge, Geneva. This exquisite timepiece is a culmination of two years of research and development. What makes this timepiece so extraordinary is the range of meticulous artisan techniques used in production. The watch is fabricated in 950 platinum and embellished with fine engravings. Functionally the timepiece features complications like a minute repeater mechanism and a perpetual calendar. The skeletonized dial is highly distinctive, with refined details like a slimline chapter ring and silver sub counters. At the heart of the watch is a bespoke mechanical movement. The caliber is chamfered, polished, and beveled entirely by hand.

Chopard

The name Chopard is renowned for the manufacture of fine jewelry and timepieces. Originally the company was founded in 1860 by Swiss watchmaker Louis-Ulysse Chopard but is now privately owned by the Scheufele family.

L.U.C Full Strike - White Gold

The L.U.C Full Strike was originally launched at the end of 2016 and was the highlight of the 20th-anniversary celebrations of Chopard Manufacture. This was the brand's first-minute repeater watch and featured several world exclusives. For example, the timepiece strikes hours, quarters, and minutes on sapphire gongs. This watch won the coveted Aiguille d'Or at the 2017 Grand Prix d'Horlogerie de Genève awards. What makes this timepiece so appealing is the spectacular multilayered dial showcasing elements of the mechanism. Other highlights include a stylish chapter ring with gold Roman numerals and delicate hour/minute hands. Beneath the luxurious 43.5mm 18-karat rose gold case lies a highly sophisticated in-house manufacture movement.

L.U.C Perpetual Twin

The L.U.C Perpetual Twin is an elegant 43mm × 11.47mm chronometer certified modern dress watch. Chopard states this model offers "gentleman connoisseurs a new expression of artful style and horological finesse." What makes this watch so alluring is the blue satin-brushed sunburst finished dial. This color has been achieved by a specialist galvanic treatment. Other details include rhodium-plated markers and hour/minute hands. Functionally the timepiece features hours, minutes, small seconds, date, and perpetual calendar indications. Powering the watch is a twin-barrel self-winding mechanical movement. The Calibre L.U.C. 96.22-L comprises 29 jewels and oscillates at a frequency of 28,800 vibrations per hour.

Christiaan van der Klaauw

In the world of horology, Christiaan van der Klaauw Astronomical is synonymous with the production of exclusive handmade astronomical watches. The company was originally founded in the Netherlands in 1974.

Planetarium

The CVDK Planetarium contains the smallest mechanical planetarium in the world. This amazing characteristic displays the orbits of Mercury, Venus, Earth, Mars, Jupiter, and Saturn around the sun in real time. As well as functional, this creation is also an elegant dress watch. Dimensionally the timepiece has a traditionally sized 40mm stainless steel case that could potentially appeal to the unisex market. Key features include an Aventurine glass dial depicting a star-filled sky and white rhodium-plated indexes. The watch is also presented on a high-quality black leather strap with a logo-engraved steel folding clasp. Beneath the stylish façade lies a twin-barrel mechanical automatic movement. The Calibre CVDK7386 comprises 35 jewels and has a power reserve of 96 hours.

Real Moon Tides

The CDVK is an exceptionally elegant dress watch with a refined contemporary appearance. According to the brand, this time-piece features the world's most accurate mechanical 3-D moon phase indication. This remarkable characteristic is perfectly showcased on the shimmering blue Aventurine dial. Other fine details include Christiaan van der Klaauw signature hour/minute hands and raised silver Roman numerals. There is also a magnificent tide indication at 12 o'clock. As a perfect finale, the timepiece is equipped with a 40mm polished steel case and is presented on a black leather strap with a logo-engraved steel folding clasp. At the heart lies a twin-barrel mechanical auto-matic movement. The Calibre CVDK7383 comprises 35 jewels and has a power reserve of 96-hours.

Chronoswiss

Chronoswiss is an independent watch manufacturer in Lucerne, Switzerland.
The brand is the brainchild of Gerd R. Lang, who prior to establishing this
company had worked for many other brands in the industry.

Open Gear Resec Paraiba

The Open Gear Resec Paraiba is designed to make a bold statement at any formal event. Chronoswiss has experimented with varying hues inspired by the semiprecious Paraiba tourmaline, unique to the Brazilian region. What makes this watch so enticing is the handmade soft turquoise and petrol blue guilloché dial. Other features include off-center hours (featuring "open gear" design), central minutes, and retrograde seconds. Dimensionally the watch is well proportioned and has a satin-brushed steel case (composed of 17 parts) measuring 44mm × 13.35mm. Beneath the contemporary façade lies a Swiss-made mechanical automatic movement. The Calibre C.301 comprises 33 jewels and oscillates at a frequency of 28,800 vibrations per hour.

CHRONOSWISS
Tourbillon
Atelier Lucerne
Limited 02/15
SWISS MADE

Open Gear Tourbillon

The Open Gear Tourbillon is one of the brand's most exclusive watches to date and is a limited edition of 15 pieces. This timepiece is also a departure for Chronoswiss, which normally makes classical style models. What makes this watch so unusual is the CVD coated 44mm stainless steel case finished in electric blue. This eclectic characteristic complements the multilayered hand-guillochéd and engraved blue dial. Other details include blue lacquered hands, off-center hours (12 o'clock), central minutes, and a majestic exposed flying tourbillon (6 o'clock). Powering the watch is a partially skeletonized mechanical hand-winding movement. The Calibre C.303 comprises 23 jewels and oscillates at a frequency of 28,800 vibrations per hour.

Corum

Corum is a luxury brand probably best known for its "Admiral Cup" series of watches. The company is based in La Chaux-de-Fonds, Switzerland, and is owned by Hong Kong–based Citychamp Watch & Jewelry Group Limited.

Golden Bridge Rectangle

The Golden Bridge Rectangle is based on the original Golden Bridge, which was rectangular in shape. For this model, Corum created an art deco–influenced 18-karat rose gold case that measures 29.50mm × 42.20mm. This complements the beautifully engraved, geometric, 18-karat gold movement showcased on the open-worked dial. The manual-winding Calibre CO 113 comprises 19 jewels and oscillates at a frequency of 28,800 vibrations per hour. Six Roman numerals surround the exposed mechanical structure. Essentially this perfectly interacts with the shapes, rivets, and materials that reflect the architecture of a bridge. As a perfect final touch, the watch is presented on a brown alligator strap with 18-karat gold triple-folding clasp buckle.

LAB 02

According to Corum, the LAB 02 is a perfect example of aesthetic perfection and extraordinary craftsmanship. Dimensionally the watch exudes generous proportions and has an 18-karat rose gold case measuring 45mm × 13.40mm. What makes this unique 1/1 piece so remarkable is the sensational skeletonized dial. Every element of the sophisticated mechanism is showcased perfectly, including the exposed flying tourbillon. Other features include a gold engraved chapter ring, Baton-style hour/minute hands, and rhodium-coated indexes. Overall the composition is very cohesive, and the finishing is first rate. Beneath the luxurious façade is a hand-winding movement comprising 33 jewels and oscillating at a frequency of 21,600 vibrations per hour.

Cuervo y Sobrinos

Cuervo y Sobrinos was originally founded by northern Spanish immigrant Ramon Fernandos Cuervo in 1862. The present-day brand (called CyS SA) has still maintained a Latin flair but is now based in Capolago, Switzerland.

Historiador Retrógrado

The Retrógrado is one of the most prestigious models within Cuervo y Sobrinos's highly popular Historiador collection. This watch has some really elegant details, including a cobalt blue face with a center part decorated with "Clous de Paris" engravings. There are other highlights, like an hours indication with eight pyramid-shaped indexes and Roman numerals and an applied CyS emblem. Dimensionally the watch is equipped with a classically proportioned steel case measuring 40mm. Functionally the timepiece features hours, minutes, central seconds, retrograde day/date, and power reserve indication. At the heart of the watch is a Swiss-made self-winding movement. The Calibre CYS 6330 comprises 30 jewels and oscillates at a frequency of 28,800 vibrations per hour.

Robusto Churchill Yalta Edition

The Robusto Churchill Yalta Edition is inspired by the historic meeting of leaders Churchill, Roosevelt, and Stalin in Yalta in 1945. As well as acknowledging an enduring friendship between the UK and US, Cuervo y Sobrinos has also created a beautiful dress watch. Key features include a stylish silver "Clou de Paris" guilloché dial with 4N gold applied indexes and an applied CyS emblem. There is also a very distinctive day/date window at 6 o'clock. This entire splendor is encapsulated in a perfectly formed 43mm stainless steel and titanium case. Powering the watch is a modified Swiss automatic movement sourced from ETA. The Calibre CYS 5206 comprises 30 jewels and oscillates at a frequency of 28,800 vibrations per hour.

Czapek & Cie

Czapek & Cie is a luxury brand originally founded by Czech-born Polish master watchmaker François Czapek in 1845. The company was resurrected in 2011 by Harry Guhl, Sébastien Follonier, and Xavier de Roquemaurel.

The Place Vendôme Platinum "Lumières"

Place Vendôme is named after the Parisian square where François Czapek opened a boutique in the mid-19th century. The "Lumières" Platinum is a beautiful limited edition dress watch—15 pieces worldwide. What makes this watch so extraordinary are the exquisite hand engravings on the side of the case, meticulously executed by artist Michèle Rothen Rebetez. Other features include a magnificent open-worked Champlevé dial with a "grand feu" white enamel ring. However, the main highlight is the exposed one-minute tourbillon at 8 o'clock. At the heart of the watch is an Haute Horlogerie proprietary mechanical hand-winding movement developed by Chronode and Czapek & Cie.

Quai des Bergues Sursum Corda

The Quai des Bergues Sursum Corda is a very special, unique piece that has been created to celebrate Czapek & Cie's 175th anniversary. With a white gold case measuring 42.5mm, this watch is an ideal choice to wear with formal attire. What makes this timepiece so remarkable is the cloisonné enamel dial, featuring a world map in the center. Other fine details include a large chapter ring with hand-engraved waves, skeletonized, rhodium-plated steel hands, and a curved antireflective sapphire crystal. The watch is also presented on a matte-blue alligator strap with a complementary white gold buckle. Beneath the elegant façade lies a Swiss-made mechanical hand-wound movement. The Calibre SXH1 oscillates at a frequency of 21,600 vibrations per hour and has a power reserve of 168 hours.

D. Candaux Watch

David Candaux is a third-generation master watchmaker based in Vallee de Joux (30 miles north of Geneva), Switzerland. Before establishing his own brand he spent decades working for several renowned watchmakers.

DC7 Genesis

The DC7 Genesis is one of David Candaux's recent exclusive models that is made to order. With a case measuring 44mm, the timepiece exudes generous proportions but is still suitable to wear with formal attire. Probably the most interesting variant is a collaboration with renowned artist Mikki Saturno for "Only Watch." This is a biennial charity auction of luxury timepieces made by the finest watchmakers for research into Duchenne Muscular Dystrophy. What makes this unique 1/1 piece so extraordinary is the hand-painted dial inspired by Leonardo da Vinci's *Vitruvian Man*. Other features include refined skeletonized hour/minute hands and a biplanar flying 30° tourbillon indication at 12 o'clock. Beneath the contemporary façade lies a sophisticated hand-winding movement.

1 | 12 | 0
D.CANDAUX
LE SOLLIAT
LE COEUR & L'ESPRIT

DC 1 "The First 8"

DC 1 "The First 8" was David Candaux's debut timepiece as an independent watchmaker. As the name suggests, this exclusive watch is limited to only eight pieces worldwide. What makes the timepiece so extraordinary is the sophisticated "grand feu" enamel dial finished in 5N gold. This works perfectly with the handcrafted, flame-blued, inverted "syringe" style hands. Other features include a large central second hand and double patented invisible crown system. The star of the show is the wonderful exposed flying tourbillon carriage at 9 o'clock. At the heart of the watch is a highly sophisticated 47-jewel manual winding movement. This caliber contains 287 components and oscillates at a frequency of 21,600 rotations per hour, and it has refined skeletonized hour/minute hands and a biplanar flying 30° tourbillon indication at 12 o'clock. Beneath the contemporary façade lies a sophisticated hand-winding movement.

DeBethune

De Bethune is a high-end Swiss watch manufacturer in L'Auberson, Switzerland.
The brand has become synonymous with the production of technically brilliant timepieces
that pay homage to the great 18th-century master watchmakers.

DB27 Titan Hawk

The DB27 Titan Hawk is a modern dress watch that should appeal to a broad demographic. It is also limited to only 10 pieces worldwide and is priced around $40,000 (making it one of DeBethune's most affordable pieces). This timepiece is presented in a 43mm grade 5 titanium case, making it suitable to wear with formal attire. What makes this watch so desirable is the concentric pattern green dial, which has a refined modernistic appearance. Other fine details include contrasting white Roman numeral indexes, skeletonized hour/minute hands, and a top-mounted crown. Beneath the contemporary façade lies a bespoke self-winding mechanical movement. The Calibre AUTOV2 comprises 29 jewels and oscillates at a frequency of 28,800 vibrations per hour.

DB28 XPTIS1

The DB28 XPTIS1 is an extraordinary dress watch from DeBethune with a futuristic appearance. Dimensionally the timepiece is well proportioned and has a 43mm diameter case that makes it ideal to wear with a dinner suit. It is also exceptionally lightweight due to the polished grade 5 titanium ergonomic case. What makes this watch so appealing is the sensational architectural 3-D multilayered titanium dial. This fine detailing perfectly interacts with all of the other polished titanium components. Other features include DeBethune patented floating lugs, an open-worked balance wheel (at 12 o'clock), and blue titanium hour/minute hands. Powering the watch is an in-house mechanical hand-winding movement comprising 194 parts.

Depancel

Depancel is a small independent watchmaker that specializes in the production of high-quality affordable timepieces. The company was founded by the dynamic Clément Meynier and is in Pont de Poitte, France.

PISTA GT

PISTA GT is probably Depancel's most exclusive watch to date and has a modern aesthetic that will appeal to a broad demographic. The brand says this is "the perfect affordable, Swiss-made skeleton watch for performance enthusiasts." Dimensionally the timepiece is presented in a well-proportioned 316L stainless steel case that measures 42mm × 10.5mm. This works perfectly with the black PVD coated bezel and black geometric skeletonized dial. Other features include a black chapter ring, bold white numerals, open-worked hour/minute hands, and an oversized crown. At the heart of the watch is a mechanical automatic movement from Swiss Technology Production. The Calibre STP6-15 comprises 26 jewels and oscillates at a frequency of 28,800 vibrations per hour.

Serie-R F-back

The Serie-R F-back is a striking dress watch with a very bold contemporary appearance. Depancel says this new watch "speaks of bold independence, defying the conventions of traditional watchmaking." With a rectangular stainless steel case measuring 43mm × 36 mm, this timepiece is ideally suited to formal occasions. Key features include a distinctive blue

dial stamped with a radiator grill design and three-color racing stripes. Other details include three sporty subdials, date indication (at 4.30), a sapphire crystal, and a perforated leather racing style watch strap. Powering the watch is a Japanese-made mechanical automatic movement. The Caliber Miyota 9120 comprises 26 jewels and oscillates at a frequency of 28,800 vibrations per hour.

F. P. Journe

F.P. Journe is a high-end Swiss watchmaker with headquarters in Geneva, Switzerland. The brand was founded in 1999 by François-Paul Journe and is the only three-time winner of the Aiguille d'Or grand prize from the Fondation du Grand Prix d'Horlogerie de Genève.

Centigraphe Souverain 10th Anniversary

The Centigraphe Souverain 10th Anniversary is a special limited edition piece (1–10) that was exclusively produced for F. P. Journe's Tokyo Boutique. This exquisite timepiece has a traditionally sized case measuring only 40mm. To add a contemporary twist, it is fabricated from polished titanium. What makes this watch so alluring is the multilayered (partially skeletonized) dial composed of ruthenium and white gold. Other interesting details include sapphire counters, 5N gold coated steel hour minute/hands, red-lacquered titanium small hands, and an 18-karat 6N gold crown. Powering the watch is an 18-karat rose gold manual-winding movement. This caliber comprises 50 jewels and oscillates at a frequency of 21,600 vibrations per hour.

Tourbillon Souverain Cadran Régence Circulaire-01

The Tourbillon Souverain Cadran Régence Circulaire-01 is an exceptionally graceful contemporary dress watch with a luxurious appearance. This particular model was the first in the Souverain collection and the first tourbillon wristwatch to feature a Remontoire (constant force device). What makes this watch so remarkable is the extravagant white gold dial with exclusive hand-engraved Régence Circulaire and silver guilloché decoration. Other refined details include stylish blue hands and a large aperture displaying the tourbillon carriage. This entire splendor is encapsulated in a beautiful 18-karat gold case. At the heart of the watch is a manual winding mechanical movement. The Calibre 1403 comprises 26 jewels and oscillates at a frequency of 21,600 vibrations per hour.

Fabergé

The House of Fabergé was originally founded by Gustav Fabergé in 1842 (Saint Petersburg, Russia) and is synonymous with highly ornate eggs. The modern incantation of the brand is in London and focuses mainly on high-end jewelry and timepieces.

Compliquée Peacock Arte

The Compliquée Peacock Arte won the Grand Prix d'Horlogerie de Genève (High-Mechanical award) in 2015. This latest addition is a collaboration with a specialist in miniaturist art, André Martinez. The watch is very small by present standards and has a demure rose gold case measuring only 38mm. These proportions should appeal to a broad demographic, including the unisex market. What makes this timepiece so stylish is the sensational hand-painted dial, depicting a peahen alongside the iconic peacock. Other refined features include a white mother-of-pearl rotating hour ring and minute track. Beneath the extravagant façade lies a Swiss-made mechanical movement. The Calibre 6901 comprises 38 jewels and oscillates at a frequency of 21,600 vibrations per hour.

Visionnaire DTZ Dynamist

The Visionnaire DTZ Dynamist is a contemporary dress watch that is made to order. Fabergé states the timepiece adopts a logical and intuitive method of displaying two time zones. Essentially this watch is ideal for people who are prolific travelers. Dimensionally this timepiece is well proportioned and has a case measuring 43mm. For this model the brand has experimented successfully with materials like black ceramic and titanium. What makes this watch so distinctive is the circular satin black DLC-treated (diamond-like carbon) chapter ring with complementary yellow gold printed numerals. Other highlights include a central GMT aperture and Fabergé signature hour/minute hands. At the heart of the watch is a bespoke 30-jewel mechanical movement developed by Agenhor.

Ferdinand Berthoud

Swiss luxury watch company Ferdinand Berthoud is named after the renowned Swiss scientist and watchmaker of the same name. The brand specializes in the production of exceptionally high quality timepieces.

Chronomètre FB 1L.4 "Far Side of the Moon"

The Chronomètre FB 1L.4 "Far Side of the Moon" is a special limited edition from Ferdinand Berthoud. According to the brand, this version evokes the dark and mysterious tones of the moon's hidden face. The octagonally shaped case is inspired by marine chronometers designed by Ferdinand Berthoud from 1760 onward. Key highlights for this piece are the luxurious 18-karat white gold case with black ceramic lugs and two transparent sapphire portholes. To complete the picture there is also a rhodium-plated sandblasted brass dial and 18-karat white gold Dynamometric crown system. Beneath the contemporary façade lies a highly complicated Swiss mechanical hand-winding movement. The Calibre FB-T.FC.L comprises 55 jewels and oscillates at a frequency of 21,600 vibrations per hour.

Chronomètre FB 1R.6-1

The Chronomètre FB 1R.6-1 is an exceptionally elegant dress watch with an avant-garde appearance. Dimensionally the timepiece is well proportioned and has a diameter of 44mm × 13.95mm. Ferdinand Berthoud has fabricated the octagonal case from ultra-strong, scratch-resistant, carburized stainless steel. This will improve long term durability and also gives the piece a modern aesthetic. What makes this watch so impressive is the innovative black rhodium-plated nickel silver dial with satin-brushed hands. Other highlights include regulator style hours presented on a rotating sapphire disc, a small counter displaying minutes, and power reserve indication. Powering the watch is a sophisticated chain-and-fusee tourbillon (constant force) mechanical movement.

Frédéric Jouvenot

Frédéric Jouvenot is a "Haute Horlogerie" Swiss independent watchmaker
that has been nominated to the Grand Prix d'Horlogerie de Genève. He has also
won several other several fine watchmaking awards.

Ace Pink Gold

The Ace Pink Gold is a beautiful 16-piece limited edition dress watch from Frédéric Jouvenot. Dimensionally the timepiece is well proportioned and has an extravagant 18-karat pink gold case measuring 44 × 55 × 13 mm. For this reason the wristwatch is an ideal choice to wear with formal attire. What makes this timepiece really stand out is the refined multilayered dial. This perfectly showcases elements of the classic column wheel chronograph mechanism. Other highlights include architectural open-worked lugs, a satin-brushed chapter ring, delicate hour/minute hands, and a black alligator leather strap. Powering the watch is a historical Swiss automatic mechanical movement. The Calibre Venus 175 oscillates at a frequency of 18,000 VPH and has a power reserve of 50 hours.

Amaterasu Titanium

The Amaterasu Titanium is part of Frédéric Jouvenot's highly acclaimed Solar Deity collection. This watch has been inspired by the great ancient sun gods worshiped by different civilizations around the world. What makes this timepiece so striking is the three-dimensional architectural multilayered dial. Other features include a central rotating disc displaying minutes and skeletonized lugs. This entire splendor is encapsulated in a 44mm matte titanium case that gives the watch a modernistic aesthetic. Ultimately this makes it exceptionally lightweight and suitable to wear with a dinner suit. Powering the watch is a sophisticated in-house manufactured hand-winding mechanical movement. The caliber oscillates at a frequency of 21,600 VPH and is decorated with Côtes de Genève engravings.

Furlan Marri

Furlan Marri is a Geneva-based watch brand that offers a range of stylish, affordable dress watches. The company is the brainchild of Swiss industrial designer Andrea Furlan and Middle Eastern artist Hamad Al Marri.

Mare Blu

The "Mare Blu" is one of Furlan Marri's latest releases that is designed for discerning clients. This particular watch is very well priced and retails for approximately $600. Dimensionally the timepiece displays very modest proportions and has a brushed 316L stainless steel case measuring 38mm. What makes the watch so attractive is the blue and purple dual-tone circular brushed dial. This works in perfect harmony with the raised silver Roman numerals. Functionally the timepiece features hours, minutes, seconds chronograph with asthmometer (five respirations) scale, and 24-hour indication. The watch is also water resistant to a depth of 50 meters. Beneath the stylish façade is a Japanese Seiko VK64 mechaquartz movement with a mechanical module.

FURLAN MARRI
Ref.1041-A

Mr. Grey

The "Mr. Grey" is an exceptionally stylish dress watch with a refined modernist appearance. Impressively, this timepiece won the "Revelation Prize" at the prestigious Grand Prix d'Horlogerie de Genève 2021 awards. With a brushed 316L stainless steel case measuring 38mm, the watch displays unisex proportions. What makes the timepiece so appealing is the matte gray sandblasted dial with raised polished indexes. Furlan Marri states they pay particular attention to the legibility of the dials, with typefaces chosen for their lightness and uniqueness. Other details include circular-grained silver counters and a dome sapphire crystal. Powering the watch is a Japanese-made Seiko VK64 mechaquartz movement with an additional mechanical module.

GENUS

GENUS is a Swiss Haute Horlogerie (or Swiss fine watchmaking) brand in Geneva, Switzerland. The company is the brainchild of watchmaker Sébastien Billières and entrepreneur Catherine Henry.

GNS Dragon

The GNS Dragon is one of the most exclusive timepieces that GENUS has produced to date. If you are searching for a dress watch with a modern twist, this model is the ideal choice. Dimensionally the timepiece has a well-proportioned, luxurious, 18-karat white gold case measuring 43mm × 18.8mm. What makes this watch so desirable is the exquisite skeletonized dial. Hours are displayed on 12 satellite indexes (peripheral and axial rotation), and minutes transition between two counterrotating discs. There is also a wonderful 18-karat rose gold stylized dragon, which is meticulously hand-engraved. Powering the watch is a manual winding mechanical movement manufactured at the brand's Geneva workshop. The Calibre 160W-1.2 comprises 26 jewels and oscillates at a frequency of 18,000 VPH.

GNS1.2

The GNS1.2 is the first production timepiece from GENUS and was originally unveiled in 2019. Impressively, this timepiece was awarded the "Exception Prize" at the Grand Prix d'Horlogerie de Genève (GPHG) in 2019. For this watch the brand has fabricated the 43mm × 13.1mm case from ethical 18-karat white gold that is RJC certified (Responsible Jewelry Council). Highlights include a sensational three-dimensional, multilayered dial with satellite hours and minutes displayed on a skeletonized disc. Other features include a dome sapphire crystal and engraved crown. Beneath the sophisticated façade lies an in-house manufacture hand winding movement. The Calibre 160W-1.2 comprises 26 jewels and oscillates at a frequency of 18,000 vibrations per hour.

Girrad Perregaux

Girard-Perregaux SA was originally established by goldsmith Jean-François Bautte in 1791 and became renowned for a historic tourbillon with three bridges. Since 2011 the brand has been owned by the French luxury group Kering.

La Esmeralda Tourbillon

La Esmeralda Tourbillon draws its inspiration from the historic Tourbillon with Three Gold Bridges pocket chronometer. This iconic watch won the gold medal at the Universal Exhibition in Paris in 1889. The modern interpretation is presented in an extravagant 44mm 18-karat pink gold case. What makes this watch so enticing is the distinctive open-worked minimalistic dial. Ultimately this perfectly showcases the architecture of the "Three Bridges" mechanism. Other details include "dauphine"-type hands and an antireflective sapphire crystal. Functionally the watch features hours, minutes, small-seconds, and tourbillon indication. Beneath the refined façade lies a 37-jewel self-winding movement that contains 310 parts and oscillates at a frequency of 21,600 vibrations per hour.

Free Bridge Infinity Edition

The Free Bridge Infinity Edition is a modern dress watch that was released to coincide with Geneva Watch Days. Girard-Perregaux's CEO, Patrick Pruniaux, states, "With this model, our master watchmakers have reimagined the company's famous Bridges." Dimensionally the watch is well proportioned and has a 44mm × 12.20mm black DLC (diamond-like carbon) treated steel case. What makes this timepiece so attractive is the modernistic, matte-black, multilayered dial. Other highlights are suspended pink gold indexes (luminescent treated) and gold "dauphine"-style hour/minute hands. At the heart of the watch is a self-winding mechanical movement with a pink gold oscillating weight. The Calibre GP01800-1170 comprises 23 jewels and oscillates at a frequency of 28,800 vibrations per hour.

Greubel Forsey

Greubel Forsey is a luxury Swiss watchmaker that is synonymous with highly expensive and complicated timepieces. The company was originally launched in 2004 by Robert Greubel and Stephen Forsey and is in La Chaux-de-Fonds, Switzerland.

Balancier S

Balancier S is an exceptionally complicated timepiece from Greubel Forsey with a striking modernistic appearance. For this model the brand has chosen to encase this watch in grade 5 titanium rather than platinum or gold. The timepiece appears circular from above but has a unique arched ovoid shape from other angles. What makes this watch so remarkable is the three-dimensional architectural dial. This reveals elements of the mechanism and features an oversized exposed 12mm balance wheel. Other highlights include skeletonized hour/minute hands, an engraved bezel, and a curved sapphire crystal. Powering the watch is a sophisticated in-house manual-winding movement. The caliber comprises 42 jewels and oscillates at a frequency of 21,600 vibrations per hour.

12
GF
QP H/M
QUANTIEME PERPETUEL
BIDIRECTIONNEL
JAN
MON
28
EQUATION DU TEMPS
TOURBILLON 24 SEC.
SWISS MADE
24 HRS

QP à Équation

QP à Équation is an ultra-complicated timepiece with its tourbillon and a complete equation of time perpetual calendar function. The overall concept is inspired by the mechanisms incorporated in age-old astronomical clocks with calendar indications. What makes this particular piece so striking is the sumptuous brown, multilayered dial. This perfectly showcases the perpetual calendar indication and exposed tourbillon carriage. Other refined details include skeletonized hour/minute hands, applied Arabic numerals/batons, and power reserve indication. Beneath the luxurious 43.5mm 5N red gold case lies a bespoke hand-winding mechanical movement composed of 624 parts. The caliber comprises 72 jewels and oscillates at a frequency of 21,600 vibrations per hour.

Grönefeld

Grönefeld describes themselves as the "Horological Brothers" and are part of a watchmaking family dating back over a century. The modern-day company is owned by Bart and Tim Grönefeld.

1941 Remontoire

The 1941 Remontoire is a multi-award winning modern dress watch with a lot of panache. This watch won the prestigious Grand Prix d'Horlogerie de Genève award (best watch in the "Men's Watch" category). For this special version, Grönefeld has created a bespoke version for a discerning client. The watch has a unique eye-catching blue fish-scale guilloché dial with multiple layers of transparent enamel to enhance the pattern. This amazing feature is created on a century-old machine. Other fine details include raised indexes, a small second counter, and extended Lancette hour/minute hands. Powering the watch is an in-house manufacture mechanical movement. The Calibre G-05 comprises 36 jewels and oscillates at a frequency of 21,600 vibrations per hour.

Decennium Tourbillon

The Decennium Tourbillon is a special limited edition timepiece (10 pieces) created to celebrate the company's 10th anniversary. Grönefeld states that "this timepiece is endowed with an array of features designed to indulge the desires of watch connoisseurs." With a luxurious platinum case measuring 39.5mm × 10.5mm, this timepiece exudes classical proportions. However, the Sterling Silver 925 matte gray lacquered dial gives the watch a more contemporary appearance. Other interesting aspects of the design include polished hour markers, an engraved crown, and a large aperture (at 6 o'clock) displaying the tourbillon carriage. Beneath the exquisite façade lies a highly complicated self-winding mechanical movement composed of 259 individual parts.

Hajime Asaoka

Hajime Asaoka is one of Japan's premier independent watchmakers who fastidiously crafts every component of his timepieces. He is a member of the elite AHCI (Academie Horlogere des Createurs Independants organization).

"Chronograph"

The "Chronograph" is inspired by Hajime Asaoka's love of traditional chronograph watches dating to the 1950s and '60s. These historical timepieces featured column wheels, horizontal clutches (with carrying arm), brake levers, and sliding gears. For this watch, Hajime has created a special skeletonized dial to perfectly showcase these elements. The level of finishing is extraordinary, and no detail has been overlooked. This entire splendor is encapsulated in a demure 38mm polished steel case. Other refined details include an oversized exposed 15mm balance wheel and blue steel hands. At the heart the watch is an entirely custom-made mechanical movement. The caliber has a power reserve of 40 hours and oscillates at a frequency of 18,000 vibrations per hour.

Project -T Tourbillon

Project -T Tourbillon is one of the most exclusive watches Hajime Asaoka has released to date. The timepiece was created in cooperation with specialist aerospace industry manufacturer Yuki Precision. This timepiece features 13 ball bearings, including one for the tourbillon carriage, which is the smallest in the world. Essentially, ball bearings improve endurance and resistance to shock. The timepiece also has two modules that maintain more stability and accuracy. Other highlights include a sensational abyss black dial with complementary skeletonized steel hands. Beneath the 43mm polished steel case lies a sophisticated bespoke mechanical movement. The caliber has a power reserve of 40 hours and oscillates at a frequency of 18,000 vibrations per hour.

Holthinrichs Watches

Holthinrichs Watches is an independent Dutch watch brand launched in 2016. The company is the brainchild of Michiel Holthinrichs, who has an obsessive passion for modern architecture and a desire for fine watchmaking.

Ornament 1 18-Karat Rose Gold

Ornament 1 18-Karat Rose Gold delicately combines modern technology and traditional craftsmanship. This timepiece is the brand's flagship model and is created for purists. The watch displays unisex proportions and measures 38mm × 7.4mm. The slimline silhouette makes this timepiece an ideal choice to wear with formal attire. What makes the timepiece so appealing is the luxurious hand-polished, 3-D printed, 18-karat rose gold case. This characteristic works in perfect harmony with the stylish ivory color enamel dial. Other highlights include open-worked hour/minute hands and a small second counter at 6 o'clock. As a perfect finale the watch is presented on a high-quality nubuck leather strap. Beneath the refined façade is a modified Swiss-made mechanical movement from Peseux.

RAW Bronze

Holthinrichs Watches says the "RAW Bronze model goes to the maximal extent of 'Horlogerie Brut' by introducing patina for extreme material textures and the physical expression of the notion of time." With a diameter measuring only 38mm, this watch exudes classical proportions. What makes this timepiece so distinctive is the textured, 3-D printed, aged bronze case with hand-polished detailing. This involved more than 100 hours of manual labor executed by the finest crafts-men. Other highlights include a handmade patined copper sunburst pattern dial with exposed copper embossed indexes. Powering the watch is a modified Swiss-made mechanical movement from Peseux. The Calibre HW-S01A features skele-tonized bridges and black polished screws.

Hublot

In the scheme of things Hublot is a relatively young watch brand and was established by Carlo Crocco in 1980. Within a few decades the company (now owned by LVMH) has grown exponentially and has luxury boutiques worldwide.

Big Bang Integral Tourbillon Full Sapphire

The Big Bang Integral Tourbillon Full Sapphire is an exceptionally glamorous dress watch designed to make a bold statement. Sapphire crystal pieces have infiltrated into the world of horology in the last few years, and this is a fine example. This piece is smaller than a lot of Hublot's other models and has a diameter of 43mm. For that reason I feel this watch would be ideal to wear to a gala ball. What makes this timepiece so remarkable is the extravagant polished sapphire crystal case, bezel, and bracelet. These elements perfectly complement the skeletonized tourbillon dial and open-worked hour/minute hands. Powering the watch is an in-house mechanical automatic movement. The Calibre HUB6035 is composed of 243 individual parts and oscillates at a frequency of 21,600 vibrations per hour.

Big Bang Sang Bleu II King Gold

The Hublot Sang Bleu II limited edition chronograph is a fusion of "Haute Horlogerie" and abstract art. For this model, Hublot has joined forces with self-described tattooist Maxime Plescia-Büchi and his design agency Sang Bleu. With a case size of 45mm × 16.5mm, this timepiece displays generous proportions. However, due to the geometric silhouette, the watch works surprisingly well with formal attire. What makes this timepiece so interesting is the way that time is presented. The open-worked dial is composed of a latticework of revolving discs, rather than conventional hands. Beneath the luxurious satin-finished and polished 18-karat King Gold case is a refined self-winding movement, the Calibre HUB1240 UNICO chronograph flyback mechanism with column wheel.

Jaquet Droz

Jaquet Droz is an exceptionally high-end watch brand in La Chaux-de-Fonds, Switzerland. The company pays homage to late 18th-century watchmaker Pierre Jaquet-Droz and is owned by the Swatch Group.

Charming Bird

The Charming Bird by Jaquet Droz is an exception in the world of watchmaking. This is the first and only timepiece to feature a singing bird. For that reason, this amazing watch was awarded the prestigious 2015 GPHG (Grand Prix Horlogerie de Genève) prize the year of its launch. Jaquet Droz says, "This automation embodies the genius of Pierre Jaquet-Droz." The watch is available in a choice of red gold and white gold and limited to only eight pieces. What makes the timepiece so desirable is the exquisite ivory Grand Feu enamel dial with platinum Roman numerals. Powering the watch is a refined self-winding single barrel mechanical movement. The Calibre Jaquet Droz 615 comprises 29 jewels (+ singing bird action, 15 jewels) and oscillates at 21,600 vibrations per hour.

Skelet-One Tourbillon Sapphire

The Skelet-One has been one of Jaquet Droz's bestselling watches that has resonated with a younger audience. Recently the brand has introduced a new exclusive model to the range called the Skelet-One Tourbillon Sapphire. Dimensionally the watch has a slightly larger diameter, measuring 42mm. Impressively, this is the world's first timepiece with a sapphire case composed of only three elements, achieved by gluing the middle and back together. Essentially this creates a minimalistic aesthetic that is devoid of screws. Other highlights include a striking skeletonized dial with exposed silicon escapement and titanium tourbillon cage. Functionally the watch features hours, minutes, seconds, and a power reserve of seven days.

Jaeger-LeCoultre

Jaeger-LeCoultre is one of the most renowned luxury watch manufacturers in the industry and was founded by Antoine LeCoultre in 1833. The company is in Le Sentier, Switzerland, and has been owned by Swiss luxury group Richemont since 2000.

Master Grande Tradition Grande Complication Rose Gold

The Master Grande Tradition Grande Complication is a beautiful astronomical dress watch from Jaeger-LeCoultre. This timepiece showcases an extraordinary level of mechanical engineering. In fact, the watch combines two of the most technically challenging complications in horology: a minute repeater and a celestial vault. What makes this timepiece so amazing is the multiple-level celestial-theme dial with an intricate structure that echoes the pattern of the constellations. Other highlights include a large aperture (at 6 o'clock) showcasing a majestic orbiting tourbillon. Beneath the luxurious 45mm rose gold case lies a refined mechanical movement. The Calibre 945 comprises 50 jewels and 570 individual parts.

Reverso Tribute Enamel Hokusai

Reverso Tribute Enamel Hokusai is a rare treasure that will be cherished for life. The Reverso was originally launched by Jaeger-LeCoultre in 1931 and is now considered by many to be a 20th-century classic. This particular piece pays homage to the art deco styling of the original model. What makes this watch so exquisite is the miniature enamel picture depicting Japanese artist Katsushika Hokusai Kirifuri's *Waterfall* woodblock print. This delightful feature was executed by a master craftsman and is showcased on the reverse of this Reverso. Other features include a sublime guilloché wave pattern dial with faceted appliqué hour markers and dauphine hands. Powering the watch is a hand-winding mechanical movement that oscillates at a frequency of 21,600 vibrations per hour.

KEFA

KEFA is the only watch brand to secure an exclusive concession from Vatican authorities to produce 2013 limited watches with the emblem of the Holy See and coat of arms of Pope Francis. The company was established in 2020 and is in the heart of Rome in Italy.

Petrus

Aesthetically the Petrus is an exceptionally stylish dress watch designed to make a bold first impression. What makes this timepiece so alluring is the meticulously engraved silver dial showcasing the coat of arms of the Vatican city-state. This heraldic emblem depicts keys surmounted by the papal tiara. Other features include raised silver indexes, delicate hour/minute hands, a sapphire crystal lens, and a small date window (at 3 o'clock). Overall the composition is well conceived, and the level of finishing is superlative. Beneath the 39mm polished steel case is a self-winding movement from manufacturer Swiss Technology Production (STP). The Calibre STP1-11 comprises 26 jewels and oscillates at a frequency of 28,800 vibrations per hour.

"Petrus for Her"

Following on the success of the Petrus, KEFA recently unveiled a ladies dress watch called the Petrus for Her. This elegant timepiece shares a lot of the same characteristics as the male/unisex version, but there are subtle differences that brand owners Carmelo Caruso and Gianluigi Di Lorenzo were keen to highlight. For example, the hour/minute hands and case have become more curvaceous, creating a more feminine aesthetic. Clients can also select from a broad range of colorful straps. To guarantee anticounterfeiting, the timepiece is equipped with a digital electronic identity card together with a serial number (engraved on the case). Powering the watch is a 26-jewel Swiss mechanical automatic movement that oscillates at a frequency of 28,800 vibrations per hour.

Konstantin Chaykin

Konstantin Chaykin is Russia's most famous watchmaker, who has created many extraordinary timepieces. Based in Moscow, he is also the president of the acclaimed Académie Horlogère des Créateurs Indépendants organization.

Minotaur

The Minotaur is a very eclectic modern dress watch that celebrates the year of the bull. For this timepiece, Konstantin Chaykin has been inspired by Chinese mythology and in particular the zodiac 12-year cycle, where the bull is ranked second. Unusually, Konstantin chose iron bronze for the 42mm case, which is a departure from other models. What makes the watch so appealing is the brown dial with guilloché relief décor, which is meticulously finished with multilayers of fine lacquer. Hour and minutes are presented on two revolving discs. There are large double-disc days of the week indications positioned at 6 o'clock. Beneath the striking façade is a Swiss-made self-winding mechanical movement. This caliber is based on the Vaucher VMF 3002 base, with a bespoke module made by Konstantin Chaykin.

KONSTANTIN CHAYKIN
12
9
3
6
60
45
15
30
U
S
E
G
N
I
K
M
O

Mouse King

Ernst Goffman's timeless classic "The Nutcracker and the Mouse King" inspired Konstantin Chaykin to create a watch of the same name. Interestingly, the conception of the watch began during the Christmas holidays, when Konstantin chose to read a bedtime story to his daughter. The end result is a beautiful limited edition (eight pieces) timepiece ideally suited to formal occasions. Key highlights include an intricate silver guilloché wave dial finished with a multicolor lacquer coating. This charming feature is encapsulated in a 42mm titanium case with an 18-karat gold crown. At the heart of the watch is a modified Swiss-made mechanical self-winding movement from Vaucher. The Calibre K07-1 comprises 36 jewels and oscillates at a frequency of 28,800 vibrations per hour.

Krayon

Krayon is an ultra-high-end Swiss watch brand that produces exceptionally expensive timepieces for discerning collectors. The owner of the company, Rémi Maillat, describes himself as a watchmaker, engineer, and mathematician.

Anywhere

Rémi Maillat describes Anywhere as an ideal of beauty, a marriage of pure elegance and legibility. Certainly this 2020 model from Krayon is a stylish dress watch with a contemporary twist. Dimensionally the timepiece is small by comparison to many other pieces available on the market. This watch has a luxurious 18-karat gold case measuring only 39mm. For that reason it will appeal to gentlemen with smaller wrists, or potentially the unisex market. What makes this timepiece so appealing is the sensational hand-painted Métiers d'Art dial. This charming detail is a modern miniature reinterpretation of Claude Monet's painting *Impression, Rising Sun*. At the heart of the watch is a bespoke 55-jewel manual-winding mechanical movement that contains 430 individual parts.

Everywhere

Everywhere is Krayon's most exclusive watch to date and definitely the most intricate. Impressively, this timepiece won the "Best Watchmaking Innovation of the Year" award at the prestigious 2018 Grand Prix d'Horlogerie de Genève. Rémi Maillat states the inspiration behind this watch is the desire to know sunrise and sunset times at any point on the globe. With an extravagant 18-karat white gold case measuring 42mm × 11.70mm, this timepiece is well proportioned. What makes the watch so exclusive is the decorated white gold dial, which displays sunrise and sunset times on peripheral discs. Beneath the elegant façade lies a highly complicated mechanical self-winding movement with a 22-karat gold microrotor. The caliber comprises 85 jewels and 595 components.

Lebois & Co

Lebois & Co is a Netherlands-based fine watchmaker that was originally founded in 1934 by Raymond Dodane. As a tribute to his legacy, Tom Van Wylick reignited the brand in 2014 and has now created many beautiful timepieces.

Airain Type 20

Montres Airain is a historic watch brand that was originally founded by the Dodane family in 1934. This company became synonymous with the production of high-quality timepieces throughout the 1950s and '60s. Type 20 is a Lebois & Co tribute to one of their most popular historic models. This reedition watch has a classical-size stainless steel case measuring only 38mm × 14mm. Ultimately these demure proportions make the watch ideal to wear to formal occasions. For this project, Tom Van Wylick collaborated with freelance watch designer Matthieu Allégre to reimagine the aesthetic of the original. Main highlights include a beautiful black dial with contrasting silver numerals and counters. Powering the watch is a 24-jewel manual-winding column-wheel chronograph movement.

CoLAB Heritage Chronograph

CoLAB Heritage Chronograph is a unique dress watch that has a great backstory. As well as creating elegant timepieces for Lebois & Co, Tom Van Wylick has also created a collaborative initiative called CoLab. This subbrand unites watch enthusiasts from around the world to cocreate new watch models. The first model is a reissue of one of their most iconic 1940s chronographs. Since Louis Moinet invented the movement more than 200 years ago, chronographs have become very popular in the world of horology. What makes this watch really attractive is the stylish salmon dial with tachymeter and telemeter indications. Beneath the 39mm stainless steel case lies a 24-jewel manual-winding column-wheel chronograph movement.

Lederer Watches

Lederer Watches is the brainchild of veteran watchmaker Bernhard Lederer, who was one of the founding members of Académie Horlogère des Créateurs Indépendants organization. He is renowned for creating exceptionally high quality timepieces.

Central Impulse Chronometer

The Central Impulse Chronometer won the "Innovation Prize" at the prestigious Grand Prix d'Horlogerie de Genève (GPHG) 2021 awards. This magnificent timepiece has taken Bernhard Lederer 10 years of fastidious research and development. The result is an incredible dress watch that is sure to resonate with connoisseurs. What makes this timepiece so remarkable is the sophisticated blue-patterned dial with two open-worked small seconds counters. This fine detail perfectly exposes elements of the refined mechanism. Functionally the watch features hours, minutes, seconds, special escapement, and a double 10-second remontoire. Beneath the 44mm 18-karat gold case lies a highly complicated hand-winding movement that oscillates at a frequency of 21,600 VPH.

BLU MT3 Tourbillon

The MT3 (Majesty Tourbillon) was released under the label BLU, which stands for Bernhard Lederer Universe. Even though it is from one of Bernhard's previous collections, it is a sensational contemporary dress watch with a futuristic appearance. It is an exceptionally ambitious timepiece that features a technically revolutionary movement containing a triple tourbillon. The three orbiting tourbillons take center place on the architectural open-worked dial. This entire splendor is perfectly encapsulated in a highly polished 18-karat white gold case. Powering the watch is a manually winding mechanism that has been entirely designed and crafted in house. The Calibre BL 707 comprises 32 jewels and oscillates at a frequency of 18,000 vibrations per hour.

Louis Erard

The history of Louis Erard dates to 1929, founded by Louis Erard and André Perret. Today the company is in Le Noirmont, Switzerland, and is synonymous with producing high-quality watches.

Excellence Guilloché Main

Excellence Guilloché Main is a stylish dress watch with a very unusual modern twist. This timepiece is limited to only 99 pieces worldwide and continues Louis Erard's odyssey in artistic craftwork. Dimensionally the watch displays traditional proportions and has a polished steel case measuring 42mm. What makes this timepiece so interesting is the matte-black trompe-l'oeil pattern dial. This feature has traditionally been hand-guillochéd by specialist manufacturer Fehr & Cie SA. Other fine details include signature fir tree hour/minute hands in blued steel and a beveled crown. At the heart of the watch is a Swiss-made mechanical automatic movement from Selita. The Calibre SW200 comprises 26 jewels and oscillates at a frequency of 28,800 vibrations per hour.

Le Régulateur Louis Erard × Alain Silberstein II

In 2021 Louis Erard unveiled a very exciting collection called Le Triptyque Louis Erard × Alain Silberstein. Le Régulateur Louis Erard × Alain Silberstein II is one of the distinctive models within the range. It is the second time the brand has collaborated with the legendary designer. This creation is a superb dress watch that should appeal to appreciators of modernistic design. With only 178 timepieces released worldwide, this limited edition sold out very quickly. What makes this watch so striking is the contemporary micro-blasted grade 2 titanium and polished grade 5 titanium 40mm case. This refined detail works perfectly with the matte-black dial with opaline (matte silver) counter (12 o'clock). Other features include white/yellow/red indexes and Alain Silberstein signature hands.

Louis Moinet

Jean-Marie Schaller is a creative powerhouse and visionary who has a penchant for fine watchmaking. His brand Louis Moinet was named after the legendary French horologist, sculptor, and painter who famously invented the chronograph movement.

Memoris Yucatán

Jean-Marie Schaller states that the "Memoris Yucatán epitomizes the immense potential of mechanics without compromising on style." Certainly this is a beautiful contemporary dress watch that should appeal to purists. The concept of this timepiece pays homage to the world's first chronograph, invented by Louis Moinet in 1816. With a grade 5 titanium case measuring 46mm, this watch displays generous proportions. Main highlights include a striking skeletonized dial that perfectly showcases elements of the mechanism. Other features include a vibrant Yucatán green chapter ring and subdial displaying hours/minutes. Powering the watch is a Swiss-made automatic movement containing 302 parts. The Calibre LM86 comprises 34 jewels and oscillates at a frequency of 28,800 vibrations per hour.

LOUIS • MOINET

Space Revolution

Space Revolution is one of Jean-Marie Schaller's finest achievements that was exceptionally technically challenging to produce. For that reason he is proud to state it is "one of the most important super watches of the decade." Not surprisingly, this exceptional timepiece was nominated for the prestigious Grand Prix d'Horlogerie de Genève (GPHG) 2021 award. The watch features two high-tech titanium spaceships and two satellite tourbillons rotating around the black Aventurine dial. This gives the timepiece a wonderful three-dimensional appearance that is sure to entice serious collectors. Beneath the luxurious 43.5mm 18-karat rose gold case is a highly sophisticated 470-part mechanical hand-winding movement. The Calibre LM104 comprises 56 jewels and oscillates at a frequency of 21,600 VPH.

Louis Vuitton

To many people Louis Vuitton is the epitome of luxury. The French fashion brand specializes in extravagant accessories like jewelry, leather jackets, shoes, sunglasses, watches, etc. and is now part of the LVMH Group.

Tambour Carpe Diem

The Tambour Carpe Diem took two years of extensive research and development to create this amazing one-of-a-kind watch. Louis Vuitton was rewarded with the "Audacity Prize" at the Grand Prix d'Horlogerie de Genève (GPHG) 2021 awards. Dimensionally this model exudes generous proportions and has a luxurious pink gold case measuring 46.8mm × 14.42mm. What makes this timepiece so extraordinary is the hand-painted enamel dial depicting a three-dimensional skull and snake illustration. Functionally the watch features hours, minutes, power reserve indication, retrograde minutes, and jumping hours. Beneath the contemporary façade lies a sophisticated hand-winding mechanical movement. The Calibre LV 525 oscillates at a frequency of 21,600 vibrations per hour.

MINUTES TOURBILLON
LOUIS VUITTON
REPETITION
SWISS MADE

Voyager Minute Repeater, Flying Tourbillon

For the Voyager Minute Repeater, Flying Tourbillon, Louis Vuitton has combined two exceptional complications in one sensational modern dress watch. For this timepiece the brand has created a contemporary satin-brushed, 18-karat white gold cushion shaped case. The case measures only 42mm × 9.7mm, making it one of the slimmest Swiss watches available on the market. Key highlights include a beautiful transparent sapphire crystal glass dial with transferred horizontal lines. This perfectly showcases the intricate mechanical elements of the timepiece. Powering the watch is a manual winding movement developed and assembled at La Fabrique du Temps Louis Vuitton. The Calibre LV100 comprises 29 jewels and oscillates at a frequency of 21,600 vibrations per hour.

Marco Lang

Marco Lang is a fifth-generation watchmaker who operates from a workshop in Dresden, Germany. For many years he operated a successful company called Lang & Heyne before establishing his own atelier in 2020.

Zweigesicht Dial 1

The concept of the first model in the "Zweigesicht" series is pretty simple but highly effective. Essentially, the future wearer has the opportunity to choose their favorite face by interchanging the strap attachment. "Dial 1" has a clean, minimalistic appearance that should appeal to appreciators of modern design. Interesting details include a white silver-plated, three-part dial with complementary Roman numerals. Other features include an applied golden five-minute motif, guilloché decorated middle section with "Clous de Paris" engravings, and pencil style hour/minute hands finished in blued steel. Beneath the 40mm steel case is a handcrafted mechanical hand-winding movement. The Calibre Ml-01 comprises 27 jewels and oscillates at a frequency of 21,600 vibrations per hour.

Zweigesicht Dial 2

"Dial 2" has a three-dimensional architectural appearance that will entice purists. What makes this timepiece so special is the white-fired skeletonized dial in solid silver. This gives the recipient a perfect glimpse of the intricate mechanics of the watch. Other fine details include a sophisticated translucent enamel "grand feu" chapter ring with blued steel, cathedral style hour/minute hands. Beneath the 40mm steel case is a handcrafted hand-winding movement. The Calibre Ml-01 comprises 27 jewels and oscillates at a frequency of 21,600 vibrations per hour.

MB & F

MB&F (Maximilian Büsser & Friends) is a high-end horological platform established by Maximilian Büsser in 2005. Essentially, the brand reinterprets traditional, high-quality watchmaking into three-dimensional kinetic sculptures.

LM101

The Legacy Machine 101 series was initially launched in 2014 and is Maximilian Büsser's interpretation of the dress watch. For the 2021 editions the brand has experimented with a selection of vibrant colors. My personal favorite is the 18-karat white gold version with a purple dial plate. This watch is one of the smallest pieces the brand has ever released and has a case measuring 40mm × 16mm. What makes this timepiece so outstanding is the multilayered architectural dial with a visible 14mm flying balance wheel. Other amazing details include a refined enamel subdial displaying hours/minutes and power reserve indication. Powering the watch is a 23-jewel three-dimensional horological mechanical movement entirely developed in-house by MB&F.

LMP Evo Blue

The Legacy Machine has proven to be one of the most successful collections for Maximilian Büsser. There have been numerous editions, including the LM Perpetual EVO, which is a collaboration with acclaimed watchmaker Stephen McDonnell. This timepiece is a lot smaller than other previous models and has a case measuring 44mm × 17.15mm. For this watch, MB&F has experimented with contemporary materials like zirconium and titanium. The overall aesthetic still makes it an ideal choice to wear with formal attire. What makes this watch so special is the highly complex skeletonized dial encapsulated in a dome sapphire crystal. Beneath the stylish façade lies a sophisticated 41-jewel hand-winding movement containing 581 individual parts.

MeisterSinger

MeisterSinger is a prestigious brand synonymous with high-quality one-handed watches. The company has won multiple design awards (including the coveted Red Dot) and is in Münster, Germany.

Bella Hora

Bell Hora is an exciting modern dress watch from MeisterSinger that has revived a practically forgotten complication called the "Sonnerie au passage." This innovative timepiece produces a friendly chime indicating the passing of each hour without having to look at the dial. For that reason, the watch won a prestigious Red Dot award. Within the range there are two dial choices: blue sunburst and natural ivory, depending on your preference. These two variants come with contrasting numerals and the brand's signature singular hour/minute hand. Beneath the 43mm stainless steel case is a modified Swiss mechanical hand-winding movement from Selita. The Calibre SW200 comprises 26 jewels and oscillates at a frequency of 28,800 vibrations per hour.

MEISTERSINGER
Pangaea
12
11
10
09
08
07
06
05
04
03
02
01
28

Pangaea Date Blue Dégradé

Pangaea Date Blue Dégradé is a superb dress watch from MeisterSinger with a refined contemporary appearance. This watch is inspired by historical wristwatches and pocket watches. For that reason, the brand has equipped this model with a scaled-down 40mm stainless steel case. These proportions should appeal to a broad demographic, including the unisex market. What makes this timepiece so attractive is the Blue Dégradé effect dial. This feature perfectly interacts with the bold silver numerals. Other details include a large date indication (at 6 o'clock), large central hand, and domed sapphire crystal. Powering the watch is a Swiss mechanical hand-winding movement from Selita. The Calibre SW200 comprises 26 jewels and oscillates at a frequency of 28,800 vibrations per hour.

Muse Swiss Art Watches

Muse Swiss Art Watches is a boutique watchmaker in St Cergue, in the canton of Vaud, Switzerland. The brand reflects the artistic aspirations of the three founding members, who have extensive industry experience.

Stella S

Stella S is a beautiful dress watch that has a high-quality artistic appearance. This timepiece has an extra-small case measuring only 37mm, which should appeal to traditionalists. To add a contemporary twist, Muse has opted for titanium in preference to stainless steel. What makes this watch so extraordinary is the black onyx dial with contrasting silver markers. Hours, minutes, and seconds are displayed on three inter-weaving satin-brushed discs which have a delightful floral pattern. This makes reading the time a very interesting experience. Powering the watch is a Swiss-made mechanical self-winding movement. The Calibre MU 01-A1 comprises 25 jewels and oscillates at a frequency of 21,600 vibrations per hour.

Tanoura

Tanoura is one of Muse Swiss Art Watches' premium models and is designed to make an exceptionally bold statement. Dimensionally the watch exudes generous proportions and has a diameter of 44mm. For this model the brand has created a polished titanium case with a refined architectural appearance. This interesting characteristic works in perfect harmony with the sparkling night sky Aventurine dial. Other features include three geometric revolving discs displaying hours, minutes, and seconds. The watch also has a rhodium plated chapter ring with satin-finished silver indexes. Beneath the modernistic façade lies a Swiss-made mechanical self-winding movement. The Calibre MU 01-A1 comprises 25 jewels and oscillates at a frequency of 21,600 vibrations per hour.

Parmigiani Fleurier

Parmigiani Fleurier SA is a luxury Swiss watchmaker that was established in 1996 by Michel Parmigiani. The company's headquarters are in Fleurier, Switzerland, and they are most renowned for creating exquisite timepieces.

Tonda PF Annual Calendar

The Annual Calendar is part of a new series of watches from Parmigiani Fleurier called the Tonda PF. This collection of timepieces features some of the brand's most distinctive complications from the outset. Dimensionally the watch has a well-proportioned 18-karat rose gold case measuring 42mm. This fine detail works perfectly with the matching 18-karat rose gold bracelet. Main highlights include a magnificent gray "guilloché Grain d'orge" dial with a large moon phase indication (at 6 o'clock). Other features include a knurled bezel, delta-shaped hour/minute hands, and 18-karat rose gold indexes. Powering the watch is a Swiss-made mechanical automatic movement. The Calibre PF339 comprises 32 jewels and oscillates at a frequency of 28,800 vibrations per hour.

Tonda PF Micro-Rotor

Parmigiani Fleurier describes the Tonda PF Micro-Rotor as an unusual and exclusive combination. Functionally the watch features hours, minutes, and date. Interestingly, the date disc is the exact same color as the minute track, creating a consistent colorway. With a polished steel case measuring only 40mm, the timepiece is small by modern standards. For that reason these reserved proportions will potentially appeal to gentlemen who prefer smaller models or the unisex market. The gray "guilloché Grain d'orge" dial and skeletonized hands give the watch a minimalistic appearance. Beneath the stylish façade lies a refined Swiss mechanical automatic movement with a full platinum micro-rotor. The Calibre PF703 comprises 29 jewels and oscillates at a frequency of 28,800 vibrations per hour.

Patek Philippe & Co.

Patek Philippe is considered the platinum standard in the world of horology and is renowned for creating exclusive timepieces. The company was originally founded in 1839 and is in the canton of Geneva, Switzerland.

5088/100P Volutes and Arabesques

The 5088/100P Volutes and Arabesques is an extraordinary timepiece from Patek Philippe designed for connoisseurs. This phenomenal timepiece exudes classical proportions and has an ultra-slim 38mm × 8.3mm case. For this model the brand has opted to use platinum, giving the watch a luxurious appearance. What makes this watch so remarkable is the hand-engraved dial, in black enamel and 18-karat gold. Other features include the stylized hour/minute hands and a shiny black alligator strap with a complementary hand-engraved prong buckle. Powering the watch is an in-house manufacture self-winding mechanical movement. The Calibre 240 comprises 27 jewels and oscillates at a frequency of 21,600 vibrations per hour.

TOURBILLON
PATEK PHILIPPE

5303R Minute Repeater Tourbillon

5303R Minute Repeater Tourbillon is a sensational dress watch that will appeal to serious collectors. Patek Philippe states that this new grand complication was created explicitly for connoisseurs of striking watches. This timepiece is presented in an extravagant 18-karat 42mm rose-gold case that features white-gold intarsia engraved with foliage motifs. What makes the watch so desirable is the majestic transparent skeletonized dial with small second counter. Other fine details include black open-worked hour/minute hands and printed numerals. At the heart of the watch is a sophisticated mechanical hand-winding tourbillon movement. The Calibre R TO 27 PS/252 comprises 28 jewels and oscillates at a frequency of 21,600 vibrations per hour.

Pilo & Co

Pilo & Co is an independent watchmaker established by Amarildo Pilo in 2001. The company is renowned for creating high quality timepieces and is in the Saint-Gervais district of Geneva, Switzerland.

DIATTO COMPETIZIONE

For the DIATTO COMPETIZIONE, Pilo & Co, Genève, has drawn inspiration from a legendary 19th-century Italian automobile manufacturer. This timepiece has a striking contemporary appearance that would work well with formal attire. Dimensionally the watch is well proportioned and has a black PVD-treated steel case measuring 43mm. What makes this timepiece so appealing is the refined skeletonized dial. This characteristic works perfectly with the black/red open-worked hands and large red markers. Other features include an oversized red crown and a high-quality black leather strap with a folding butterfly clasp. At the heart of the watch is a Swiss-made, 23-jewel, mechanical self-winding skeletonized movement with Côtes de Genève decoration.

PILO & Cº
GENÈVE
SAT SU
FRI THU WED
POWER RESERVE
SWISS MADE

Pieces d'Exception P0604HAS 1

The P0604HAS 1 is a beautiful modern dress watch from a collection called Pieces d'Exception. This series is probably the most exclusive offering from Pilo & Co, Genève. With a stainless steel case measuring 44mm × 12mm, this timepiece displays generous proportions. However, the slim profile of the watch still makes it suitable to wear with a dinner suit. What makes this timepiece so elegant is the multilayered dial with contrasting blue counters and chapter ring. Functionally the watch features hours, minutes, seconds, day, date, power reserve, and GMT indications. Beneath the stylish façade lies a high-quality Swiss-made self-winding movement. This Caliber comprises 25 jewels and oscillates at a frequency of 28,800 vibrations per hour.

Rajamäen Kellotehdas

Rajamäen Kellotehdas was founded in 2007 and specializes in the production of stylish limited edition timepieces. The company is the brainchild of watchmaker Simo Ylitalo and is in Rajamäki, Finland.

12M

The 12M is the debut model from Rajamäen Kellotehdas and is available in many different variants. Simo Ylitalo's primary objective with this model was to create a timeless dress watch with a contemporary appearance. Dimensionally the timepiece exudes traditional proportions and has a polished steel case measuring 41mm. What makes this watch so special is the refined three-dimensional, multilayered gold dial. This characteristic has an architectural aesthetic that should appeal to appreciators of modern design. Other features include stylized hour/minute hands and a high-quality leather strap with a complementary steel buckle. Powering the watch is a Swiss-made mechanical automatic movement. The Calibre ETA 2896 comprises 22 jewels and oscillates at a frequency of 28,800 VPH.

18M

The 18M is the second model released by Rajamäen Kellotehdas and can be customized to the client's specifications. With a hardened stainless steel case measuring 41mm this watch is designed to last. What makes this timepiece so attractive is the three-dimensional architectural blue and red anodized aluminium alloy dial. Hours and minutes are displayed via two discs that are visible through 24 laser-cut windows. Other features include stylized triangular hour/minute hands that are mounted on a central sapphire crystal disc. Beneath the contemporary façade lies a modified Swiss-made mechanical automatic movement. The Calibre ETA 2896 comprises 22 jewels and oscillates at a frequency of 28,800 VPH.

Raymond Weil

Raymond Weil is a high-quality independent watchmaker that was established by Raymond Weil and Simone Bédat in 1976. The company is now under the helm of Elie Bernheim and is in Geneva, Switzerland.

Freelancer 2785

With the Freelancer 2785, Raymond Weil states that they combine materials from the past with Swiss precision craftsmanship. This has proven a very successful model for the company because of its refined aesthetic. Even though the watch has a classical 42mm diameter, the appearance is contemporary. The case is fabricated from steel and the bezel from bronze. This combination of materials is proving to be very popular in modern watchmaking. What makes this timepiece so striking is the matte-brown skeletonized dial with Perlage decoration. Other features include a brown chapter ring with contrasting white numerals and a bronze crown. Powering the watch is a mechanical self-winding in-house movement, with a visible balance wheel.

THE
BEATLES
RAYMOND WEIL
GENEVE
SWISS MADE

Maestro Beatles "Let It Be"

Maestro Beatles "Let It Be" is Raymond Weil's tribute inspired by the legendary band's last album in 1970. This timepiece has an elegant appearance that should appeal to a broad demographic. Dimensionally the watch displays modest proportions and has a 40mm unisex stainless steel case. What makes this timepiece so appealing is the brushed anthracite dial with a UK-shape aperture. This gives the recipient a subtle view of the mechanism. Other features include the Beatles logo, baton-shaped indexes, and a vegan leather strap. At the heart of the watch is a high quality Swiss automatic movement. This caliber is perfectly showcased via the sapphire crystal exhibition case back.

RPaige watch

RPaige Watch is a boutique watch brand based in picturesque Honolulu, Hawaii.
The company is the brainchild of fourth-generation watchmaker Richard Paige, who
formerly owned several high profile watch stores in California.

"Holy Grail"

"Holy Grail" is a very special and unique 1/1 piece that has a beautiful appearance. RPaige specializes in sophisticated dress watches, and this model is a great example. With a polished steel case measuring 44mm × 14mm, the timepiece is well proportioned. What makes this watch so enticing is the exquisite 10-karat, antique, four-tone gold, guilloché decorated dial. This detail works in harmony with the raised Roman numerals and blued steel, Breguet style hour/minute hands. For this timepiece, Richard Paige has selected an exclusive Waltham "Riverside" Caliber manufactured in Massachusetts in 1904. This 19-jewel mechanical hand-winding movement features a rose gold balance and oscillates at a frequency of 18,000 vibrations per hour.

Military Dial Speakeasy

The 44mm stainless steel case of the Speakeasy was inspired by the glorious art deco design era. Essentially this was reflected in architecture (Empire State Building, Chrysler Building), art (Picasso), and fashion. Every model within the limited series of 50 has a unique dial. For this watch, Richard Paige has opted to use a black military style dial with antique style Arabic numerals. Other interesting details include a static vintage "escape wheel" (at 12 o'clock) and chrome cathedral-style hands. The timepiece is also presented on a high-quality handmade ostrich leg strap with contrasting white and blue stitching. At the heart of the watch is a Swiss Elabore-grade mechanical hand-winding movement. The Calibre Unitas 6497 comprises 17 jewels and oscillates at a frequency of 18,000 VPH.

Sartory Billard

Sartory Billard is a small independent boutique watch brand that specializes in customized pieces for discerning clients. The company is the brainchild of Armand Billard, who is a self-confessed artist, designer, and visionary.

SB04

SB04 is the debut model from Sartory Billard and has a crisp, minimalistic appearance that should appeal to a broad demographic. Effectively, Armand works with customers to create the desired aesthetic. Ultimately, clients can select from a range of different options for the dial. For example, stone, meteorite, and guillochés seem to be very popular at present. There is also the choice of a classically proportioned 40mm × 11.5mm polished steel or titanium case. For this model the composition is uncluttered, with a clean ivory enamel dial with contrasting black Roman numerals. Powering the watch is a mechanical automatic movement from Swiss Technology Production. The Calibre STP3-13 comprises 26 jewels and oscillates at a frequency of 28,800 vibrations per hour.

SB05

Armand Billard states the "SB5 would not be possible without the strides made with the SB04." Certainly this model is marketed at a higher price and offers clients more options than its predecessor. For that reason, the brand has collaborated with Comblemine, who creates the guilloché dials, and the case is manufactured by Voutailinen & Cattin SA. With a diameter of 38.5 mm (available in steel, titanium, gold, or tantalum), the watch is very small by modern standards. What makes this piece so attractive is the frost-gray dial with contrasting black enamel chapter ring. This interacts perfectly with the white numerals and polished steel hands. Beneath the elegant façade is a 21-jewel mechanical hand-winding movement manufactured in Switzerland by La Joux Perret.

Schwarz Etienne

Schwarz Etienne is a Swiss luxury brand established by Paul Arthur Schwarz
and his wife, Olga Etienne, in 1902. The modern-day company is in La Chaux-de-Fonds,
Switzerland, and proudly manufactures all their timepieces in house.

Roswell 08

Roswell 08 is an exceptionally high-quality contemporary dress watch with a refined modernistic appearance. The Roswell collection draws inspiration from the mystery surrounding the Roswell, New Mexico, UFO site. This is reflected in the 45mm stainless steel case, which takes design cues from the dynamic curves of a flying saucer. What really makes this watch so appealing is the multilayered, open-worked tobacco color dial with exposed elements of the mechanism. Other features include silver rhodium treated indexes and green date indication. The timepiece is also presented on a hand-sewn grained leather strap. Powering the watch is an in-house manufacture automatic movement with micro-rotor. This caliber comprises 36 jewels and 191 individual components.

La Chaux-de-Fonds Tourbillon PSR

Tourbillon Petite Seconde Rétrograde Schwarz Etienne is from the prestigious La Chaux-de-Fonds collection. This timepiece takes full advantage of the brand's capabilities to produce everything in-house. Housing the watch is a well-proportioned 44mm stainless steel case measuring 44mm × 13.70mm. These dimensions make it an ideal choice to wear on formal occasions. Key highlights include an exquisite skeletonized dial with rhodium treated hands. There is also an off-center green Aventurine fine stone subdial displaying hours and minutes (at 5 o'clock). Beneath the stylish façade is a sophisticated mechanical self-winding tourbillon movement. The Calibre TSE 121.00 comprises 40 jewels and oscillates at a frequency of 21,600 vibrations per hour.

Speake-Marin

Speake-Marin SA was established by acclaimed English Watchmaker Peter Speake-Marin SA in 2002. The company relocated its headquarters from Bursins to Geneva in 2019 and is renowned for exclusive mechanical timepieces.

Minute Repeater Flying Tourbillon Légèreté

The Minute Repeater Flying Tourbillon Légèreté is part of the Haute Horlogerie collection and has a striking appearance. Speake-Marin states that they "push the limits of the craftsmanship to keep reinventing exclusive pieces that haven't been seen elsewhere." This amazing creation is a unique 1/1 piece that has a generously proportioned sapphire crystal case measuring 46.4 mm. What makes this watch so outstanding is the multilayered skeletonized dial. Other refined features include a large aperture displaying the flying tourbillon carriage and heart-shaped blued steel hands. Beneath the modernistic façade lies an in-house manual-winding mechanical movement. The Calibre SMAHH-02 oscillates at a frequency of 21,600 vibrations per hour and has a power reserve of 72 hours.

TOURBILLON
OPENWORKED
FIVE & TWO
SPEAKE-MARIN

One&Two Openworked Flying Tourbillon

Speake-Marin enjoyed a lot of success with "Openworked Hours & Minutes" and "Openworked Dual Time" watches. Therefore it was a natural progression to release another model called the Openworked Flying Tourbillon to the One&Two collection. This timepiece is limited to 10 pieces and comes in a choice of 38mm or 42mm 5N gold cases. What makes the watch so impressive is the skeletonized dial with matte-black microblasted chapter ring. This perfectly showcases intricate elements of the complicated mechanism. Other features include gold heart-shaped hour/minute hands with SuperLuminova and a power reserve indication (at 7:30). At the heart of the watch is a self-winding movement with a micro-rotor and 60-second flying tourbillon.

TRIBUS

In the scheme of things, TRIBUS is a relatively young watch brand based in Liverpool, England. The company is the brainchild of industry veteran Christopher Ward and his three sons.

TRI-04

TRIBUS describes the TRI-04 Power Reserve GMT Sport COSC as a watch engineered for traveling in style. This timepiece is capable of tracking two time zones simultaneously and is designed for people who like to travel. With a gunmetal PVD-treated case measuring 41mm × 12.09mm, these modest dimensions should appeal to the unisex market. What makes this watch so appealing is the matte-black dial with contrasting red GMT counter at 6 o'clock. Other highlights include raised, polished gunmetal indexes and power reserve indication. Powering the watch is a Swiss mechanical automatic (chronometer certified) movement from Soprod. The Calibre C115 comprises 31 jewels and oscillates at a frequency of 28,800 vibrations per hour.

TRI-06 "My Dog Sighs"

For the TRI-06 "My Dog Sighs" collection, TRIBUS has teamed up with UK-based artist Paul Stone (a.k.a. "My Dog Sighs"). The brand says the concept was to create "50 individually hand-painted dials, each bearing the fingerprint of the unorthodox and innovative artist." Each model is different, and there is nothing else similar on the market. Dimensionally the timepiece exudes classical proportions and has a black PVD-treated stainless steel case measuring 41mm. This model also has a double-domed sapphire crystal lens that pays homage to the dress watches of the '40s and '50s. Beneath the contemporary façade lies a Swiss mechanical automatic movement from Selita. The Calibre SW200 comprises 26 jewels and oscillates at a frequency of 28,800 vibrations per hour.

Trilobe

Trilobe is a very new watch brand that was established in 2018 by the dynamic Gautier Massonneau. The company is in the provincial city of Paris, France, and specializes in avante-garde style watches.

Les Matinaux

The inspiration behind the conception of the Les Matinaux is a poetry collection by René Char. Gautier Massonneau has created this timepiece for clients with more eclectic tastes, but he has still housed the timepiece in a classically proportioned 40.5mm 18-karat rose gold case. These dimensions should appeal to watch aficionados who regularly attend formal events. What makes this timepiece so appealing is the high-quality satin-brushed silver dial. This features counterclockwise rotating aluminium hour/minute chapter rings. There is also an open-worked small second counter that draws inspiration from the rose windows of Sainte-Chapelle (a Gothic chapel) in Paris, France. Beneath the contemporary façade lies a modified ETA 2892 movement manufactured by Jean-François Mojon.

Nuit Fantastique Secret

The design of the Nuit Fantastique Secret has a minimalist aesthetic and plays with perceptions. Trilobe says, "It defies symmetrical shapes and is made up of four parts, three of which are in permanent rotation." For that reason, the timepiece won the Grand Prix d'Horlogerie de Genève nomination (in the category "Petite Aiguille"). Dimensionally the watch exudes modest proportions and has a polished steel case measuring 40.5mm × 9.2mm. What makes this timepiece so amazing is the bespoke dial, which depicts a map of the sky with the stars. This can be personalized for clients to represent a time and place that is precious to them. At the heart of the watch is a Swiss-made, self-winding movement with a micro-rotor. This mechanism has a power reserve of 48 hours and oscillates at a frequency of 28800 VPH.

Ulysse Nardin

Ulysse Nardin SA is a Swiss luxury watch brand with a history dating to 1846. The company is based in Neuchâtel, Switzerland, and is synonymous with creating accurate marine chronometers and complicated timepieces.

Tourbillon Free Wheel 44mm

The Tourbillon Free Wheel is an exceptionally complicated timepiece with an "haute horlogerie" appearance. Ulysse Nardin says this is their most spectacular innovation to date, and it literally turns watchmaking upside down. This wristwatch has a well-proportioned, luxurious, 18-karat rose gold case that measures 44mm in diameter. What makes this timepiece so spectacular is the carved stone dial with seemingly floating tourbillon bridges. This amazing feature is perfectly encapsulated in a box-domed sapphire. Other fine details include a special anchor tourbillon and power reserve indicator at 4 o'clock. Powering the watch is a sophisticated in-house manufacture hand-winding movement that utilizes Silicium technology.

Grand Deck Marine Tourbillon

The Grand Deck Marine Tourbillon is a very elegant nautically inspired modern dress watch. Essentially this watch is a continuation of the brand's tradition of creating high-precision seafaring instruments since the 19th century. With an extravagant 18-karat white gold case measuring 44mm, this watch is an ideal choice to wear with a dinner suit. What makes this timepiece truly remarkable is the handcrafted wood marquetry dial. This amazing feature perfectly frames the large aperture, showcasing a 60-second flying tourbillon. Interestingly, jumping hours are displayed by a large window and minutes on a graduated arc. Beneath the stylish exterior is a Swiss-made mechanical hand-winding movement that oscillates at a frequency of 21,600 vibrations per hour.

Vacheron Constantin

Vacheron Constantin is the oldest Swiss watch manufacturer in the world and has an illustrious history dating to 1755. The company is now owned by the Richemont Group and is in Plan-les-Ouates (canton of Geneva), Switzerland.

Métiers d'Art Tribute to Great Explorers

Vacheron Constantin states that they are celebrating their spirit of adventure through a new 10-piece limited edition model called the Métiers d'Art Tribute to Great Explorers. This sensational dress watch comes neatly packaged in a luxurious 18- karat 4N pink gold case measuring 41mm × 11.68mm. What makes the timepiece so seductive is the hand-crafted Grand Feu enamel dial, which depicts selected miniaturized portions of a 1519 map from the Miller Atlas. Other interesting highlights include a revolving disc displaying minutes, and a special dragging satellite hour indication. Powering the watch is an in-house manufacture "Hallmark of Geneva" certified self-winding movement. The Calibre 1120 AT comprises 36 jewels and oscillates at a frequency of 19,800 vibrations per hour.

VACHERON CONSTANTIN
GENEVE

Traditionnelle Tourbillon

The Traditionnelle Tourbillon is a delicate fusion of "haute horlogerie" and artistic craft. This amazing watch was released in 2020 and is an exclusive 18-piece, numbered limited edition. Dimensionally the watch exudes classical proportions and has a diameter of 42mm. For this model, Vacheron Constantin has created an exquisite hand-engraved, 18-karat 5N pink gold case and bezel. What makes this timepiece so collectable is the hand-guilloché black galvanic-treated dial with hand-engraved, 22-karat, 5N pink gold applied Qilin. Other highlights include stylish sword-shaped hour/minute hands and an exposed tourbillon at 6 o'clock. Beneath the elegant façade lies an exceptional in-house manufacture mechanical hand-winding movement with a 14-day power reserve.

Von Doren

Von Doren is Norway's premier watch brand that creates distinctive, high-quality timepieces with a touch of vintage flair. The company is the brainchild of Øyvind VonDoren Asbjørnsen, who is also a film producer/director, author, and entrepreneur.

Grandmaster Caïssa
(Norway Chess Limited Edition)

Øyvind VonDoren Asbjørnsen says the Caissa Automatic (Norway Chess Limited Edition) is truly a piece of art and as individual as the man who wears it. Certainly this timepiece has a very elegant contemporary appearance that will appeal to a broad demographic. In fact, this was so popular it sold out exceptionally quickly. Dimensionally the watch displays modest proportions and has a polished steel case measuring 43mm. What makes this timepiece so appealing is the circular-brushed brass dial with its own unique pattern and patina. This detailing works perfectly with the large open-heart aperture a 12 o'clock. Powering the watch is a high-quality Swiss mechanical self-winding movement. The Calibre STP5-15 comprises 26 jewels and oscillates at a frequency of 28,800 vibrations per hour.

Grandmaster Mark II Open Heart

Grandmaster Mark II Open Heart is a beautiful limited edition watch (200 pieces) that is designed to make a bold impression. The brand says the timepiece evokes the spirit of chess which has captivated legions of followers for well over 1,000 years. Presented in a polished steel case that measures 43mm, the watch exudes classical proportions. For this timepiece, Øyvind VonDoren Asbjørnsen has created a refined engraved dial with an open-heart feature. This gives the recipient an intimate glimpse of the mechanism. Other highlights include blued steel hour/minute hands and a domed sapphire crystal. Beneath the elegant exterior is a Swiss self-winding movement. The Calibre STP5-11 comprises 26 jewels and oscillates at a frequency of 28,800 vibrations per hour.

Voutilainen

Voutilainen is an exclusive watch brand in Môtiers, Switzerland. The company is the brainchild of Finnish watchmaker and "Académie Horlogère des Créateurs Indépendants" member Kari Voutilainen.

28ti

28ti perfectly encapsulates the philosophy of Kari Voutilainen and is a timeless dress watch. The timepiece was designed, built, fabricated, finished, and assembled at Voutilainen workshops in Môtiers, Switzerland. This watch is housed in a classically proportioned titanium case measuring 39mm × 13.40mm. What makes this timepiece truly spectacular is the multilayered skeletonized silver dial. This gives the wearer an intimate viewing of many mechanical elements of the watch. Other features include gold hour/minute hands and a blue chapter ring. Beneath the stylish façade lies a sophisticated in-house manual-winding movement. The caliber comprises 32 jewels and oscillates at a frequency of 18,000 vibrations per hour.

Starry Night Vine

The Starry Night Vine combines Japanese traditional art with Swiss haute horlogerie. The result is an extraordinary dress watch with an absolutely beautiful appearance. For this unique 1/1 watch, Kari Voutilainen has teamed up with legendary Japanese lacquer maker Mr. Kitamura and Mrs. Anita Porchet. What makes this timepiece so special is the exquisite dial, with an illustration of the sky above a vineyard. The miniature artwork involved meticulous techniques of fine lacquering that took hundreds of hours to execute. This entire splendor is housed in a luxurious 18-karat Palladium 39mm gold case. Powering the watch is a refined in-house mechanical hand-winding movement. The caliber comprises 30 jewels and oscillates at a frequency of 18,000 vibrations per hour.

Wessex

Wessex Watches is in Chippenham, Wiltshire, England, and was established in 2015 by Jamie Boyd. He prides himself on being a longtime mechanical watch enthusiast, collector, and watchmaker.

Peerless Auto

The Peerless Auto is a continuation of the Wessex Peerless Collection. Each piece that Jamie Boyd creates is bespoke and entirely made to the client's specifications. This model is presented in a polished steel 42mm "cushion" style case measuring 42mm. Essentially that makes this watch an ideal choice to wear at formal events. Main highlights include a double-layered dial in solid 925 silver with intricate guilloché decorations. Other fine details include an engraved chapter ring with Roman numerals, cathedral style blued steel hour/minute hands, and a sapphire crystal lens. Beneath the elegant exterior is a self-winding mechanical movement from Swiss Technology Production. The Calibre STP1-11 comprises 26 jewels and oscillates at a frequency of 28,800 vibrations per hour.

Peerless Handwound

The Peerless Handwound is a model from the highly successful Wessex Peerless Collection. This is another unique piece that Jamie Boyd has created for a discerning client. Dimensionally the watch is housed in a well-proportioned polished steel case measuring 43mm. What makes this timepiece so attractive is the double-layered solid 925 silver dials with intricate guilloché decorations. Other features include an engraved chapter ring with Roman numerals, blued-steel hour/minute hands, and a small seconds counter (at 6 o'clock). At the heart of the watch is a Swiss-made hand-winding mechanical movement from ETA. The Unitas 6498-1 Élaboré comprises 17 jewels and oscillates at a frequency of 18,000 vibrations per hour.

Yvan Monnet

Yvan Monnet is a luxury watch brand in the heart of Geneva, Switzerland. The founder (and namesake) has an obsession with a precision that prompted him to develop a range of stylish timepieces.

FIVE

FIVE is the brand's debut watch, which has a refined minimalistic appearance. Yvan Monnet says the shaped bezel has five straight curves that intersect a circle, with sharp angles to form a pentagon. Dimensionally the timepiece is classically proportioned and has a diameter of 43mm. The pentagon-shaped case is fabricated from satin-brushed steel, giving it a contemporary aesthetic. What makes this watch so elegant is the rhodium-plated spruce-green vertical satin-finished dial with super-Luminova® treated raised white Arabic numerals. Other features include sandblasted hour/minute hands and an antireflective sapphire crystal lens. At the heart of the watch is a Swiss-made mechanical automatic movement with a customized blue rotor.

Mina

The Mina is a delicately crafted dress watch with a high-quality appearance. Yvan Monnet says the shape of this watch is immediately identifiable and combines originality with a rare elegance. With a polished steel case measuring 35mm, this timepiece displays very modest proportions. Certainly this model would appeal to gentleman who appreciate extra-small watches or the unisex market. What makes this watch so stylish is the blue lacquered dial with contrasting raised silver indexes. Other features include skeletonized hour/minute hands and an ornate crown. Beneath the sculptural exterior lies a Swiss-made mechanical automatic movement from Selita. The Calibre SW300-1 comprises 25 jewels and oscillates at a frequency of 28,800 vibrations per hour.